# TEXTBOOK OF ECONOMIC BOTANY

# TEXTBOOK OF ECONOMIC BOTANY

**Dr. Shubhrata R. Mishra**
*Dept. of Botany*
*Vikram University*
*Ujjain (M.P.)*
*(India)*

**DISCOVERY PUBLISHING HOUSE PVT. LTD.**
**NEW DELHI-110 002**

First Published - 2010
Reprinted - 2016

ISBN: 978-81-8356-589-9



**Textbook of Economic Botany**

*Published by:*
**DISCOVERY PUBLISHING HOUSE PVT. LTD.**
4383/4B, Ansari Road, Darya Ganj
New Delhi-110 002 (India)
*Phone*: +91-11-23279245, 43596064-65
*Fax*: +91-11-23253475
*E-mail*: discoverypublishinghouse@gmail.com
sales@discoverypublishinggroup.com
*web*: www.discoverypublishinggroup.com

*Printed at:*
Infinity Imaging Systems
Delhi

# Preface

Plants are essential to the balance of nature and in people's lives. Green plants, i.e., those possessing chlorophyll, manufacture their own food and give off oxygen in the process called hotosynthesis, in which water and carbon dioxide are combined by the energy of light.

Plants are the ultimate source of food and metabolic energy for nearly all animals, which cannot manufacture their own food. Besides foods (e.g., grains, fruits, and vegetables), plant products vital to humans include wood and wood products, fibres, drugs, oils, latex, pigments, and resins. Coal and petroleum are fossil substances of plant origin. Thus plants provide people not only sustenance but shelter, clothing, medicines, fuels, and the raw materials from which innumerable other products are made. People use plants for things other than food. Plants also provide fiber, which is the tissue of plants from the stem, leaves, seeds or roots Plants in the Environment.

Plants provide many useful drugs. Some of these plants have been used as medicines for hundreds of years. The bark of the cinchona tree was used 400 years ago to reduce fever. It is still used to make quinine, a drug used to treat malaria and other diseases. Another drug, called digitalis, is used in treating heart disease. It is made from the dried leaves of the purple foxglove plant. The roots of the Mexican yam are used in producing cortisone, a drug useful in treating arthritis and a number of other diseases.

Plants play the most important part in the cycle of nature. Without plants, there could be no life on Earth. They are the primary producers that sustain all other life forms. This is so because plants are the only organisms that can make their own food. Animals, incapable of making their own food, depend directly or indirectly on plants for their supply of food. All animals and the foods they eat can be traced back to plants.

**– Author**

# Contents

# 1 Introduction

Close to 2.5 billion years ago, the earth's surface and atmosphere were stable enough to support primitive life. Single-cell organisms began to develop in the seas that covered the planet. A simple organism known as blue-green algae appeared and spread across the seas. Blue-green algae used sunlight and water to make food, and in the process, created oxygen. As the blue-green algae grew in the earth's seas, they began to fill the atmosphere with oxygen. The oxygen that blue-green algae produced made it possible for other types of organisms to develop.

Plants play the most important part in the cycle of nature. Without plants, there could be no life on Earth. They are the primary producers that sustain all other life forms. This is so because plants are the only organisms that can make their own food. Animals, incapable of making their own food, depend directly or indirectly on plants for their supply of food. All animals and the foods they eat can be traced back to plants.

The oxygen we breathe comes from plants. Through photosynthesis, plants take energy from the sun, carbon dioxide from the air, and water and minerals from the soil. They then give off water and oxygen. Animals and other non-producers take part in this cycle through respiration. Respiration is the process where oxygen is used by organisms to release energy from food, and carbon dioxide is given off. The cycles of photosynthesis and respiration help maintain the earth's natural balance of oxygen, carbon dioxide, and water.

Leaves are the main food-making part of most plants. They capture energy from sunlight, and turn water and carbon dioxide into sugar and starch. This sugar and starch becomes the food that provides plants with energy to grow, to produce flowers and seeds, and carry on their other life processes.

## Plant Facts

Scientists believe there are over 260,000 species of plants. Some plants are so small they can barely be seen. Others are taller than people or animals. One of the largest living plants on the earth are the sequoia trees of California. Some stand over 290 feet (88 metres) high and measure over 30 feet (9 metres) wide. Certain characteristics of plants set them apart from other living things.

Both plants and animals are complex organisms that are made up of many types of cells, but plant cells have thick, rigid walls that consist of a material called cellulose. Animal cells do not have this material. The cellulose enables plants to stand upright without the aid of an internal or external skeleton.

## Plants and their Environment

Plants require a reasonable level of heat to grow. The most favourable temperature at which photosynthesis takes place ranges from near freezing to 20 to 25°C (70 to 80°F). The rates of photosynthesis and respiration increase with rising temperatures. Any temperatures above or below these levels limit plant growth. The climate of a region determines what types of plants can survive in that region. A plant's environment is made up of many factors. One of the most important is the weather—sunlight, temperature, and precipitation (rain, melted snow, and other moisture).

Soil and other plants and animals that live in the same area are also included in the environment of a plant. All these factors form what is called a natural community. No two natural communities are exactly alike, but many resemble one another more than they differ. Botanists divide the world into biomes—natural communities of plants, animals, and other organisms.

## Medicine

Plants provide many useful drugs. Some of these plants have been used as medicines for hundreds of years. The bark of the cinchona tree was used 400 years ago to reduce fever. It is still used to make quinine, a drug used to treat malaria and other diseases. Another drug, called digitalis, is used in treating heart disease. It is made from the dried leaves of the purple foxglove plant. The roots of the Mexican yam are used in producing cortisone, a drug useful in treating arthritis and a number of other diseases.

## Importance of Plants

Plants are essential to the balance of nature and in people's lives. Green plants, i.e., those possessing chlorophyll, manufacture their own food and give off oxygen in the process called hotosynthesis, in which water and carbon dioxide are combined by the energy of light. Plants are the ultimate source of food and metabolic energy for nearly all animals, which cannot manufacture their own food. Besides foods (e.g., grains, fruits, and vegetables), plant products vital to humans include wood and wood products, fibers, drugs, oils, latex, pigments, and resins.

Coal and petroleum are fossil substances of plant origin. Thus plants provide people not only sustenance but shelter, clothing, medicines, fuels, and the raw materials from which innumerable other products are made. People use plants for things other than food. Plants also provide fiber, which is the tissue of plants from the stem, leaves, seeds or roots Plants in the Environment.

As a critical part of the ecosystem, plants provide oxygen for organisms to survive. They are able to reduce the problem of pollution, by using carbon dioxide. Plants are also the basis of most food webs as producers of food for herbivores and ultimately carnivores. Plants also provide shelter for animals, clean and filter water and help prevent soil erosion.

## Plants for Food

Nearly 75% of the the world's food supply is based on seven major crops: wheat, rice, maize (corn), potatoes, barley, cassava and sorghum.

## Plants for Fibre

Plants provide fibres for clothing, paper and shelter. The aboriginal people from the west coast wove cloth from the bark of the western red cedar tree. Much of our clothing today comes from synthetic (manufactured) material, such as polyester and nylon. Natural fibers also provide resources for cloth:

*Cotton* - is a natural fiber that absorbs moisture and then allows it to evaporate easily, making it the world's most important non-edible plant. The cotton fibers come from the plant's seeds. The silky fibers are strong, flexible and have a gradual spiral that causes the strands to interlock when twisted, making them ideal for spinning into thread. The second layer of fibers are shorter and are 'fuzzy' - they are used to make cotton batting, rayon and various types of plastic and paper.

*Hemp* - Early makers of jeans used hemp, which is the oldest cultivated fibe plant in the world. Other products included the Bible, sails and ropes. Hemp has a less negative effect on the environment, because it uses less land area than trees, can be harvested in a year, lasts longer than paper, can be recycled up to seven times, chokes out weeds naturally and is not prone to insect pests.

*Flax* - is a food and fiber crop. The flax fibers, which are smooth and straight, are taken from the stem of the plant are two to three times stronger than cotton fibers. Flax fiber is used for making linen paper, linseed oil - which is used as a drying oil in paints and varnish - and in products such as linoleum and printing inks.

## Plants for Medicine

An apple a day keeps the doctor away! Many medicines (over 7000) contain ingredients made from plants. Herbal remedies are a common example of how plants are used to prevent illness.

Plant medicines include:

- tea (made from ginger root) - is used to soothe an upset stomach

- white willow bark - is used to ease pain
- opium poppy's seed pod - thick milky fluid provides a powerful pain medication - morphine
- codeine is also found in the poppy - it is used in cough medicines
- Quinine - which comes from the cinchona tree - is used to prevent malaria.

### Plants for Transportation and Construction

Rubber is one of the most important plant products that people use. Natural rubber comes from the Brazilian rubber tree. Synthetic rubber is made from coal and oil by-products - but natural rubber is also an important ingredient. Canoes were carved from trees by Aboriginal people. Lubricants are provided from coconut and castor bean oils. The construction industry in North America uses wood (softwood lumber from British Columbia) as a building material.

### Plants for Fuel

Wood or coal (which is a fossil fuel) is used to heat homes. Sugar can be turned into ethanol and wood can provide methanol (wood alcohol). Fuel from plants is economical, but not energy efficient, because a large amount of energy is needed to grow the plants and a lot of the energy is lost when it is converted to fuel.

### Human Needs and Plant Needs

Our task is to make sure that plants survive and thrive in order to have this important resource in the future.

### Food

Plant parts contain carbs, fats, proteins, and minerals (calcium, magnesium, iron, etc.)

### Fruits and Veggies

- Fruit is the part of the plant that contains seeds.
- Vegetative part of the plant is any non-reproductive part.

**Root Crops**

- Potatoes.
- Rich in calories.
- Easy to grow.
- Root crops grow underground.
- Sweet potatoes,carrots, radishes, turnips.

**Legumes**

- Protein rich.
- Soybeans = most important.
- Peas, peanuts, and beans.
- Alfalfa, fed to livestock.

Cereals are the most important source of food.

- Cereals=grasses that are grown as food for humans and livestock.
- Dry fruit=grain.
- More than 1/2 of the calories humans consume come from just three species of cereal.

**Wheat**

- High in carbs.
- 1/3 of pop uses wheat as primary food source.
- Breads-pasta.

**Corn**

- Most widely cultivated crop in the US.
- Corn bread, grits.
- Chief food for farm animals.
- Corn belt= Iowa, Nebraska, Minnesota, Illinois, and Indiana.

**Rice**

- 1/2 of the world, rice is major part of their diet.

- Rich in carbs Wood.
- After food, wood is the most valuable resource obtained from plants.
- 75% in US used for building.
- Other uses:paper-source of fuel for heating and cooking.

**Medicines**

- Aspirin.
- Foxglove.
- Rosy periwinkle.

*Food.* This includes plants consumed by humans as major constituents of food preparations, and clearly comprises the most important class economically. Important specialized sources for food data include Plants for Human Consumption.

*Food additives.* Here plants are included that are also consumed, but as minor constituents of food preparations.

*Animal food.* We recognize only two subclasses: (1) fodder for plant material harvested and fed to domestic animals; and (2) forage for plant material upon which domestic animals feed themselves.

*Bee plants.* No subclasses are used for this use class. While numerous species throughout the world are of some importance as honey plants, an exhaustive treatment of these has not been attempted.

*Invertebrate food.* Only a small number of plants serving as hosts for beneficial invertebrates, such as silkworms, have been identified. Doubtless we have missed others of some importance.

*Materials.* This class includes a number of important economic plants, such as those which furnish fiber, timber, gums, resins, and industrial or essential oils.

*Fuels.* Included here with more traditional sources of fuel are rapidly growing plants with potential to provide biomass for electricity generation.

*Social uses*. This minor use class includes few plants of widespread economic importance, although tobacco and those which provide illicit drugs such as opium or cocaine are of considerable economic impact.

*Vertebrate poisons*. The economic impact of the poisonous plants classified here is largely negative. Although fish poisons provide a very important benefit in primitive cultures, they are seldom traded commercially and have been treated only superficially. Nearly all plants recorded here are toxic to humans and livestock, collectively classed as "mammals" in level 2. Only a few plants toxic to birds, chiefly poultry, have been distinguished, although most plants poisonous to mammals doubtless apply here also. Our list of poisonous plants is certainly not exhaustive, as many plants other than those treated here would certainly be toxic to humans or domestic animals if they were consumed, but this effect has not been properly documented.

*Non-vertebrate poisons*. The plants classified here provide materials that serve as organic pesticides.

*Medicines*. Here we include plants that serve as sources of specific pharmaceutical agents and those that are widely used, mostly in the crude sense, as "folklore" remedies. the numerous plants used locally as medicines have been excluded, as have those plants being tested for potential medicinal benefit. It is difficult to determine the current commercial importance of many medicinal plants, so some important ones may have been missed and others included unnecessarily.

*Environmental uses*. Many valuable economic plants are classified here, including those used for erosion control, soil improvement (i.e., green manures or cover crops), and agro forestry, but most included here are ornamentals. The ornamentals were probably the most difficult to evaluate as to whether or not to include particular species. We have tried to glean from the sources utilized only the most commercially important because an exhaustive treatment of this group is beyond our scope and is available elsewhere.

Hybridization has played a major role in the development of many improved cultivars of ornamentals, and some are of complex hybrid origin. Only those which can adequately be represented by binomial or trinomial botanical nomenclature are included. Many are best treated with cultivar or cultivar-group names, which have not been treated or accepted in this account.

*Gene sources.* In addition to known or potential sources of beneficial genes for specific improvements in crop plants, numerous species can be successfully crossed with crop species to provide fertile progeny and thus must be considered important as genetic resource plants. These primary gene pool members have been included for most crops with the designations "related to" or "progenitor of " preceding the crop name at subclass level.

Similarly, those plants that have been hybridized extensively to produce important ornamentals are labelled "for ornamental cultivars" at this level. A few species important as models for research in plant biology are included in this category. Weeds. This class is, like vertebrate poisons, one of negative economic impact. Included here are both weeds of croplands and natural habitats. Those weeds which are listed in the rules of seed-testing organisations like the Association of Official Seed Analysts and the International Seed Testing Association have been labelled "possible seed contaminant". It is difficult to judge the negative costs of some weeds, but we have attemped to include all the most troublesome ones.

*Harmful organism host.* This class includes plants that serve as alternate hosts for crop pests and plants that serve as test organisms for detecting viral diseases in crops.

Plants have many basic uses in human society, and we will look at most of them during this unit. The focus of this unit will be, however, the use of plants as food. While we use plants as a source of building materials, clothing and medicines, our survival is completely dependent upon the ability of plants to synthesize high energy molecules that we use as food.

Our economic prosperity is completely dependent upon our ability to provide food at very low cost. Since food is absolutely essential, the greater the expense of time and money food, the lesser the time and money available for all other things.

Traditionally, plants have been considered to be a pure food. Ancient references, such as *The Bible* indicates the traditional consideration of plants as the more wholesome, pure food for humans. In the paradise of Eden, Adam and Eve were sustained by food from plants. The first plants cultivated (about 10,000 years ago) were grains, like rice and wheat, as well as legumes such as beans. Other plants, such as those producing fruits and vegetables were cultivated later. Obviously fruits and vegetables were collected from the wild prior to cultivation.

Grains in particular have been subjected to selective breeding from times of earliest cultivation. We will look at the origins of wheat in particular later in the unit. Even today, grains form staples in our diet, while fruits and vegetables are usually served as side dishes.

Herbs and spices have come to be used as flavoring for foods. Two common definitions of herbs and spices can be found. One has herb as common plants and spices as tropical plants. The other, which we will use, defines herbs as leaves and stems and spices and roots, seeds and nuts.

Two other basic botanical terms are fruit and vegetable. In reality, a fruit is a flowering plant reproductive structure. The defining part of a flowering plant is, of course, the flower. The structure called the carpel is where the seeds develop. Fruits are structures that develop from a carpel, and range from coconuts to tomatoes to apples to walnuts.

A vegetable is any plant or any plant part. Technically all fruits are vegetables but not all vegetables are fruits. Plant life in general is called vegetation. As foods, we have historically referred to sweet tasting plant parts as fruits and unsweet tasting plant parts as vegetables. In this sense, grains are not considered fruits, although technically they are fruits!

The cultivation of plants is the most important activity in civilized cultures. Large scale cultivation of plants for foods by a small number of people in society allows all others to engage in other pursuits. If we each had to grow and/or find our own food, we could not maintain the diverse society we all enjoy. All of the civilized cultures in the world developed around areas that allowed for large scale cultivation of plants. The regions where large scale cultivation of plants has proven most reliable have developed into the most powerful civilizations.

Our western European culture developed in the East, where wheat has been the basis of agriculture. In the Far East, rice has been the basis of agriculture. Central and South American are regions where cultures have developed based on the cultivation of plants like potatoes and corn. As each culture developed, a greater number of plants came under cultivation, like vegetables and spices. The "discovery" of the "new world" by Columbus was driven by a desire to gain easier access to spice producing regions in the Far East. In addition to wheat, the staple, we find most of the fruits, vegetables, herbs and spices that we use today used by ancient Greeks and Romans. Plants such as figs, dates, mustard, basil, marjoram, cumin, lettuce, cherries, cucumbers and many more were commonly used in the ancient Mediterranean cultures.

By the middle ages, there were records of the use of plants such as spinach, celery, and broccoli. The first plants imported into the European diet from the Far East were plants such as apricots, peaches, cinnamon and ginger. Later plants like orange, eggplant, clove and tarragon were "discovered" and imported in the western European diet. Upon the "discovery" of the "new world", plants like potatoes, sweet potatoes, pineapples, vanilla, and kidney beans were added to the western European diet, resulting in the sort of cuisine that we enjoy in the United States today. Improved techniques of cultivation, shipping, and storage make an array of plants products available to us as food that is unmatched at any time in history, or anywhere else in the world even in the present. One of the most amazing things about our culture is the selection of food products available in supermarkets, as well as the affordability of those products. We will study the major food plants of the world, their places of origin, the places they are now grown and how they are grown.

# 2

# Agroforests as an Alternative to Pure Plantations

## INTRODUCTION

Domestication of forest species for purposes of commercial cultivation seems to have bright prospects. Forestry definitively needs to find an answer to the exhaustion of wild resources while in the same time rationalizing the production of marketable NTFPs. There is great demand for new crops and new markets. Sustainable development has to mitigate the effects of deforestation by increasing the planting of trees on cleared lands, and tropical farmers have to find substitutes for the natural resources lost through deforestation.

However, domestication is neither a fast nor an easy process. It took centuries of farmers' work to develop a range of food crop varieties and decades of scientific research to create productive clones of industrial tree crops. Can foresters or farmers wait another 50 years to tame wild rattans, or before they efficiently produce natural resins or marketable 'forest fruits' from improved and domesticated trees? Can we wait for agroforestry research to spend several more decades testing efficient tree-crop associations for forest domesticates and making them acceptable to smallholder farmers?

Modern plant domestication relies on relatively sophisticated techniques. But it should not be restricted to only a set of techniques. Domestication is also, and primarily, a strategy. Domestication techniques are selected on the basis of economic choices and by the initial preference for a given cropping system within a given farming system. In discussions on domestication, this latter aspect is often neglected, probably due to the historical background of the present scientific community. Modern agricultural and forestry science evolved in Europe, in a pastoral and cereal-growing civilization, which influenced most of the present agricultural paradigms and agroecosystem models.

Our perceptions of domestication and agricultural development are deeply influenced by a tacit preference for cereal-based models, which may not be the most useful for the domestication and development of tropical forest resources. The design of the agroecosystem in which domesticated species are grown is as essential as the choice for particular plant selection, breeding and reproduction techniques. This is particularly important when switching from annual crops to trees and from fields to forests.

Most 'modern' agroecosystems in the world relate to a single agricultural model: intensive, highly specialized stands of homogeneous crops. However, other available models exist, which should be examined under the new perspective of domestication of tropical forest species. Especially important among these are the native farming systems in the tropics, for which there is academic information already. This presentation will concentrate on the implications that ecological models in such native systems might have for the design of novel domestication strategies for NTFPs, particularly in agroforestry. An 'ecosystem' perspective to the domestication of forest species could change the approach chosen and lead to changes in techniques and processes.

It is essential to understand that the importance of an ecosystem approach to domestication goes far beyond biological or technical considerations. It also has obvious sociopolitical and institutional dimensions. The common preference of governing elites and scientists for the plantation model of domestication for commercial forest resources in the tropics did not only change the face of forest landscapes and national economies. It also deeply affected forest communities and their socioeconomic life. It is essential to analyse in the context of NTFP domestication, this indirect link between the choice of a particular domestication strategy and the fate of local populations. Domestication is part of a resource appropriation process, and resource appropriation by a powerful fraction of the population might lead to dispossession of the weaker fraction. This aspect of domestication will be examined in the light of the Indonesian history in forest resource management.

## FROM NATURAL RESOURCES TO CULTIVATED CROPS

### Food Crop Domestication and Agroecosystems

The process of domestication and cultivation of food crops has followed two totally divergent routes: 'agriculture' and 'horticulture', taken in their etymological sense-the cultivation of 'ager' (openfields) and the cultivation of 'hortus' (gardens). Several ethnobotanists and anthropologists have suggested that relating these 'grain' and 'garden' models to existing cropping systems will help to focus further discussion. 'Agriculture' refers to the 'grain model' developed for cereal domestication in ancient Mesopotamia and around the Mediterranean, and also in ancient rice civilizations from India to China.

'Ager' (literally the tilled or totally cleared field) conquered the forest and is the central platform for domestication and the 'home' of domesticates. It represents a highly specialized and artificially uniform openfield, devoted to grain crops, largely disassociated from the pre-existing environment. It involves a massive population of genetically homogenous plant cultivars. The grain model relies on highly specialized and segregated technical knowledge relating to monocultures as well as to intensive use of chemical and mechanical inputs. These are associated with very high energy consumption and minimal labour inputs at the field level. Genetic specialization and manipulation culminate in the widespread use of hybrids, which totally depend on people for their production and regeneration. The grain model perfectly relates to the productionist mentality of modern agriculture. Initially devised for food crops, it has deeply influenced modern commercial tree cropping: both tropical plantation agriculture as devised by European colonialists (e.g., rubber, oilpalm, cocoa, or cinchona) and commercial fruit culture in temperate regions. Both situations replicate a modern corn field biologically and technically, as well as in ideology.

'Horticulture' refers to the management of 'hortus', the garden, which can be characterized by high plant diversity, including tuberous perennials and trees, in somewhat chaotic combinations and configurations. The garden model involves diverse production, from many food crops. Domestication in the hortus, whether in temperate or tropical regions, has operated through the treatment of plants as individuals. It developed countless cultivars of fruits, tubers and vegetables. In the tropics, 'hortus' can be a swidden, an anthropogenic forest, or a homegarden. Management in the garden model plays intensively on ecosystem interfaces and fully benefits from natural vegetation dynamics. Devised for multipurpose production like indigenous agriculture in the tropics, as well as for an optimum management of ecological and economic risks, the garden does not comply with the urgent need for 'productivity' in agriculture. Temperate gardens are nowadays devoted mostly to production for home consumption or to leisure, whereas their tropical relatives are neglected or even denigrated by official agricultural services.

In the one-tracked mind that characterizes modern rural development, the grain model is considered as the only model that is valuable for efficient agricultural production.

Agroforestry research and extension represents the only consistent scientific framework through which the hortus model could be further developed. But this requires innovative approaches, especially as far as domestication and cultivation models are concerned. The openfield preference has often led to reductionist approaches to domestication, which we have to re-examine in the light of agroforestry development. Does domestication unavoidably come down to adapting a wild species to grow in industrial plantations? Are tropical gardens confined to the production of, at best, semi-domesticates? What new perspectives can agroforestry bring to domestication?

## Domestication: Which Models for Forestry?

Domestication and cultivation have long been the prerogative of agriculture (here in its widest sense: management of fields and gardens), while worldwide, forests remained the domain of hunting and gathering in support of agriculture. Until roughly the beginning of the 19th century, tropical and temperate forests were mainly managed for integrated, multipurpose use. These practices more or less maintained forest composition, providing for grazing, hunting, gathering, and wood or timber production. Some species selection did happen, creating useful genotypes that could be substituted for wild ones in niches within the pre-existing biotic community. Among other examples are productive chestnuts and sweet acorn varieties in the Mediterranean countries, and peach palm and brazil nuts in the Amazon basin.

Rationalization of forest production and forest culture did not really appear before the industrial era. This coincided with more focused demands for specific forest products: fuel for metallurgy, timber and pulp. The development of scientific forestry induced a strict partition between agricultural and forest development. Productivism applied to forestry created new models of intensive wood production, inspired from the grain model in agriculture, with the multiplication of highly homogeneous, specialized, and productive plantations of pine in the north and of Acacia and Eucalyptus in the tropics. Today's commercial forestry in the tropics, dominated by the imperatives of wood supply, follows a bipolar model: products are either extracted from natural, more or less managed natural stands, or cultivated in highly specialized plantations.

The monoculture preference stands as the recognized option for domestication strategies for timber trees. Domestication and cultivation techniques devised by forestry research for tree species follow the rules of specialization, uniformity and intensification that have proven efficient for grain crops, simplifying the structure and the function of the cultivated 'forest' to the extreme. But how far does this model extend to NTFPs?

What happened in the domestication and cultivation of tropical NTFPs? Domestication of non-timber forest products in the tropics developed under three different situations.

## The Indigenous Perspective

The history of management, selection and cultivation of useful forest species by indigenous people for subsistence purposes is as old as that of humankind's use of forest ecosystems. In terms of domestication, the process of interaction between farmers and forests gave rise to a whole range of modified tree varieties, which most authors classify as 'managed species' or 'semi-domesticates'. However, scientists commonly recognize that true domestication occurred through the transfer of edible forest species-primarily fruits and nuts-to agriculture through various types of garden. These forest domesticates are presently considered as true agricultural or horticultural crops.

Though not often acknowledged in the literature, commercial gathering of forest products by local people also went along with plant selection, integrated management and even cultivation. This simple domestication developed interesting models of true 'forest culture', that relate to the hortus model, and will be discussed further. These models have unfortunately remained neglected in most discussions and are almost never included in development programmes, even in agroforestry.

## COLONIAL INTERVENTIONS

Tropical NTFPs acquired a new dimension during the expansion of the colonial era. This started with the high demand for spices (nutmeg, cinnamon, vanilla), coffee or cocoa, which created a new, but highly specialized commercial perspective. These cash crops boomed with the rise of the industrial era, which needed new products, such as latexes and resins. For some products, a demand boom, the need for increased control over the resource or over product quality, led to domestication through intensive cultivation. The process of domestication and cultivation developed by the colonial plantation managers induced a true movement of disassociation between forest resources and forest ecosystems. Most of the new economic forest resources were transferred to plantation agriculture, through vast estates that reached their peak in the second half of the 19th century and the early 20th century, with oil palm, rubber, cinchona, coffee, tea and cocoa. In transferring forest resources to agriculture, the colonial plantations did not attempt to innovate; they followed the grain model, creating huge areas of specialized, artificial and productive tree monocultures.

## NEW FOCUS ON NTFPs

The commercial interest in NTFP declined after World War II, because of the fall of colonial powers and the rise of petrochemical industries, which developed substitutes for natural products. However, a new focus on NTFP exploitation has recently emerged. NTFP extraction is now considered a promising alternative to timber extraction in natural forest management. The justification for this revived interest is not entirely based on economics, although new markets for natural products have emerged, such as phytochemicals in pharmaceutical industries or wild substances in food industries, that have triggered economic interest in NTFP. The other justifications have been ecological and sociopolitical. Ecological interest has focused on reduced disturbance to the forest ecosystem *vis-a-vis* timber extraction practices-while sociopolitical interests have centred around the promotion of new development models for indigenous forest people and the promotion of 'fair trade' for natural products.

For some of the most coveted products, development through domestication and commercialization tends to emerge as a strategy competitive with extraction from natural forests. Thus, despite the debate about developing new and better models for the domestication and cultivation of useful tree species in farm lands, there is a good probability that the conventional monocultures will persist. This is already obvious for rattan production in Indonesia and Malaysia, where specialized plantations have been established under the strict control of forestry services, and for the Brazil nut in Brazil, where private investors have established large plantations that severely compete with smallholder farmers and extractivists activities.

## Consequences of NTFP domestication and commercialization for indigenous people: Lessons from the past.

The era of colonial trade and management had two major consequences for forest people: the displacement of their effective control over the collection and trade of forest resources and passage of the resources either to the governing elite or to private colonial entrepreneurs. This resulted in the local forest communities being overlooked and in the consequent transfer of the commercial benefits to private and commercial planters. This 'dispossession process' continued through most post-colonial governments and is still common today when a traditional forest resource encounters a commercial boom. This abuse of native rights commonly starts with restrictions over harvesting practices, develops with the attribution to acknowledge traders of monopolistic rights for gathering, and culminates with plantation development. This process has political support and is a practical consequence of commercial interests.

The history of natural rubbers in the Amazon and Southeast Asia perfectly illustrates this dispossession process, which transferred control over the resource-from native collectors to powerful traders and then to planters. Caoutchouc (para rubber from *Hevea brasiliensis*) was traditionally collected and used by Amerindians in the Amazon Basin. In Southeast Asia, especially Sumatra and Borneo, other wild rubbers (*Ficus elastica* and *Willughbeia sp.*) were harvested and traded by local swiddeners before the 17th century. Commercial interest for wild rubbers in Europe and the United States boomed in the second half of the 19th century, but until 1900, both elastic and non-elastic rubbers remained supplied exclusively by wild species through latex extraction. As prices rose, the control of the rubber areas in Amazonia became concentrated in the hands of the political-economic elite, who 'exerted an absolute rule over native rubber tappers'. At the same time in Indonesia, the Dutch colonial government progressively restricted exploitation by local tappers, first through imposing a licence to tap the trees, then through granting all tapping rights to foreign concessionaires. Another period of dispossession of native tappers started with the cultivation of the Amazon rubber tree in Southeast Asia in 1877. By 1913, the supremacy of the wild caoutchouc came to an end and the local tappers in the Amazon virtually stopped working. Cultivated rubber production in the Far East, through large estates, had captured the market.

Such examples are numerous, starting from the Dutch VOC taking possession of local nutmeg production and trade in the Moluccas in 1621 to the present Indonesian seizure of the birds' nests caves of the local Punan in East Kalimantan, who had owned and managed them sustainably for centuries. To get full control of the nutmeg trade, the Dutch established specialized nutmeg plantations worked by imported slaves while destroying nutmeg trees in surrounding islands. Nobody except VOC was allowed to grow and trade nutmeg any more.

### What are the prospects for domestication of forest species through agroforestry?

Forestry in the tropics is still looking for technical and economic as well as socially accepted models. Can it escape from the bipolar model extraction in natural forests on one hand and monospecific specialized plantations on the other? Can domestication of forest trees help to alleviate this situation, or will it contribute to increase the present movement of segregation between forest and agricultural development? Who will benefit from domestication of NTFP: foresters, planters, or smallholder farmers? How can rural development efficiently and sustainably incorporate the so-called minor forest resources into farm lands?

Agroforestry is often cited as the most favourable means of providing positive answers to such questions. But what type of agroforestry is the best for NTFP development? Is agroforestry research able to fully integrate and develop the potential offered by these 'new' crops? Can it generate 'new' models of agroecosystems, combining both agricultural and forest qualities in an economic as well as ecological perspective? Most agroforestry research until now has concentrated on simple associations, trying to introduce a single tree or shrub species into former grain-model systems. Most agroforestry research has promoted fast-growing trees or shrubs, not old-growth forest species. Agroforestry has been more 'agro' than 'forestry' oriented, more crop than tree based. When dealing with NTFP domestication and cultivation, agroforestry research definitely has to innovate. It has to give new preference to tree-based systems and to look for new ecosystem models (see Leakey 1996). There is room for innovative experimental research. However, a reanalysis of current management practices of forest resources by smallholder farmers might help. Can we find, in indigenous systems, management models that fully integrate forest resource management into farmlands? Can these models inspire novel domestication and cultivation strategies for forest species and thus a new phase of agroforestry development?

## NTFPs in Indigenous Agroforestry Practices: A Long History of Coevolution

Incorporating forest resources in farming systems is not a new practice in the tropics; various traditional forms of agroforestry have developed around NTFPs, constituting the very basis of indigenous agricultures in tropical America, Africa (see Boffa, this volume) and Southeast Asia. Beside fruit-dominated gardens, indigenous farmers have integrated true forest culture into farm lands, and these forms are usually closely associated with shifting cultivation. These 'cultivated forests', which often complement subsistence food cropping in annual fields, are established after total removal of the original vegetation, and they constitute complex, tree-based agroforestry systems, which fully deserve the name of 'agroforests'. Why have they been overlooked in agroforestry research? Perhaps it is a consequence of their appearance: they do not look like cultivated ecosystems and have mainly been mistaken for natural forests.

## A FOCUS ON INDONESIAN PRACTICES

Many true 'agroforests' in Indonesia have evolved around fruit and nut trees. But there are other systems that have evolved from former extractive practices in natural forests, through the deliberate incorporation of non-food forest tree species that farmers have cultivated for products to be marketed in international trade.

Among the commercial forest products, cinnamon [(*Cinnamomum burmanii* (C.G. & Th. Nees) Bl.)] was probably the first to have been incorporated into indigenous agricultural systems. In the central highlands of Sumatra, indigenous stands have been established for more than two centuries. Some form specialized, homogeneous gardens, but others, established on steep slopes, associate cinnamon trees as an understorey with higher canopy trees grown for fruits or timber. Similarly, benzoin (Styrax) is known to be managed as a fallow crop in what represents a true rotational agroforestry system in Laos. Such rotational systems were also mentioned in North Sumatra as long ago as the 18th century. However, other cultivation practices also developed through more complex and permanent agroforestry systems, which associate benzoin trees in a mix of useful timber and fruit trees.

In western Borneo, swiddeners have, for at least 150 years, established highly diversified tree gardens that integrate oil-producing dipterocarps (11 Shorea species) together with tens of other fruit and nut species as well as rattans, latex-producing trees and timber species. In Central and East Kalimantan, rattan, which has traditionally formed the bulk of trade in forest products, has been incorporated into shifting cultivation systems for more than 100 years. Rattan gardens mix the cultivated palms with planted fruit and timber trees, as well as with numerous other useful species that have established spontaneously. A century ago in the south of Sumatra, swidden farmers started cultivating damar trees (e.g., *Shorea javanica* K. & V.) for resin production and have established more than 20000 hectares of complex forest-like agroforests, associating damar with numerous other fruit and timber tree species. Native rubber trees also happened to have been planted in complex gardens, but rubber agroforestry really developed with the incorporation of the para rubber tree in local swidden systems at the beginning of this century. The Amazon rubber tree found its ecological niche in complex tree gardens in Southeast Asia, where it is grown with numerous other species, either planted or spontaneously established. It soon replaced native rubbers in the economic niche of the local swidden farmers.

All these agroforests result from farmers' needs and their deliberate choice to improve production and control, or sometimes to protect or even restore useful forest resources. They all present important common features:

1. most of them concern true old-growth forest species, not fast-growing pioneers;
2. they have all evolved from swiddens, through the systematic introduction of trees in cleared lands;
3. most often, they start as specialized plantations that evolve into a permanent mixed stand of planted tree crops and useful spontaneous resources; and
4. they exhibit forest-type structure, including a predominance of large trees, a multilayered vertical configuration and a closed-cover canopy.

Some of these indigenous systems, like the dipterocarp agroforests, hold structural as well as functional characteristics typical of a primary forest ecosystem, with the predominance of big trees, a high species richness, a high ecological complexity, and a closed nutrient cycle. Others, like the rubber agroforests that cover the lowlands of Sumatra and Kalimantan, are more like secondary forests, with dense stands of smaller trees and a rapid turnover of species. These agroforests combine important income-generating strategies based on forest resources and diversified subsistence strategies, but they are not isolated management units: they always complement other agricultural activities, such as food cropping in open fields.

Lastly, they rely on local representation and knowledge systems evolved from former forest traditions; they are maintained by simple techniques and integrated practices and are controlled by a well-defined social and tenurial system, which includes rights as well as duties.

## Ecosystem Domestication: A New Perspective?

The current concept of tree domestication appears poorly adapted to characterize and critically analyse most of the above-mentioned examples of integrating forest resources into agricultural farming systems. 'Domestication' usually focuses more on selection and propagation techniques, making too little reference, if any, to the concepts and strategies developed for the integration of wild resources direct into the farming system. This last point is nevertheless essential when dealing with the domestication of wild forest species. The preeminence of the grain model has obscured analyses from other perspectives. Is the conventional model of domestication the best choice for trees that have evolved in a highly diverse and structurally complex environment?

Usually, domestication has intentionally disassociated the resource from its natural habitat. Transfer of the candidate species to an artificially prepared environment has been seen as essential, to allow increased human control of the plant, as well as to induce and efficiently select useful genetic variations. But should the artificial cultivated environment be fundamentally different from the natural one?

Although a very artificial environment has proven its efficiency in cereal domestication, as well as in colonial tree-crops development and modern forestry plantations for wood production, there is seemingly no satisfying theoretical answer for NTFPs. The domestication strategy exhibited in the Indonesian agroforest examples partly relies on conventional plant species domestication techniques (selection, reproduction and planting practices), but it does not involve crop management in highly specialized stands, which are quite different from the original conditions in which the wild species had evolved.

Nor does it involve a major modification of the structural and biological features of the tree species, in which trees are selected to allow their adaptation to homogenous monocultural conditions. Rather, the agroforest model relies on an artificially induced reconstitution of a true forest-like ecosystem, simulating the basic principles of a natural silvigenetic succession, which allows the selected species to establish, grow and reproduce as in their original habitat.

Establishing an agroforest is conceived as a specialized tree-planting process aimed at controlling and concentrating the selected forest resource; but this process is achieved through the integration of the resource with natural vegetation and through several successional stages that lead to the gradual reconstruction of a diversified forest structure. The forest tree seedlings are introduced into the forest clearing with other short- or medium-cycle crops (e.g., rainfed paddy, vegetables, coffee bushes, pepper vines) and receive the care given to the crops. After the abandonment of these food crops as the tree canopy closes, the planted trees are strong enough to grow along with secondary vegetation and overcome competition from pioneers.

The subsequent tree fallow then freely develops with little damage to planted trees. The structure of the agroforest becomes more complex over the years, as the consequence of a particular form of management that maximizes the use of natural production and reproduction processes in order to minimize the rarest economic factor: labour. In a maturing agroforest, plant species regenerating from the neighbouring forests, through natural dispersion, can establish while forest animals find shelter and feed. Through selection, farmers favour economic resources, but non-economic resources are allowed to reproduce.

And after several decades of such a balance between free-functioning and integrated selection, the mature phase of the agroforest resembles a natural forest more than a conventional tree plantation. This ecosystem-analogy strategy has proved efficient for quick acclimation of a true forest tree species. For example, forest farmers in Sumatra have succeeded in what most foresters dream about, but have failed to achieve: the establishment, maintenance, and regeneration of a healthy dipterocarp plantation at low cost and on a huge scale.

This is a unique example for the whole forestry world. Dipterocarp agroforests rely on selected and planted forest trees; they exhibit high-density stands and good productivity, but they are also characterized by good ecological sustainability, low-cost establishment and easy regeneration over years. This is quite uncommon in conventional plantation forestry.

The agroforest domestication process allows the maintenance of biological diversity in the qualities of the tree, as it does not focus on the selection of single-purpose varieties: the multipurpose dimension of the wild species is not lost through domestication, as usually happens in the conventional process. Low-canopy varieties of durian (*Durio zibethinus* Murr.) or rambutan (*Nephelium lappaceum L.*) selected by plant breeders produce nothing else but fruits, but damar or benzoin trees domesticated for resin production by local farmers in Sumatra are still good timber producers. This multipurpose dimension is also maintained at the ecosystem level.

In the plantation model, what is not the crop is a weed. In the agroforest, self-established species are integrated as economic and ecologically beneficial resources or kept as potentially useful species. But the most original point is how natural biological processes are utilized to support the artificial domestication and cultivation process. In the agroforest, natural vegetation dynamics are channeled, first to speed up and secure the integration of slow-growing trees in the cultivated system, then to maintain a continuous balance between obsolescence and the regeneration of the cultivated stand. Beside domesticating forest species, the agroforest allows the restoration of integral biological and ecological processes, which determine the overall survival and success of the cultivated ecosystem. These natural processes schematically replace the high technology and energy inputs of forest plantations.

Here, it is the agroecosystem that adapts to the plant characteristics. In this sense, agroforests constitute an original attempt of 'ecosystem domestication', through the full utilization of natural ecosystem dynamics to the benefit of a selected, artificially established population of trees. Through this, the agroforest domestication strategy proves to be successful in assimilating the problems associated with the long-term management of forest tree species-or vines, as in the case of rattan. Long-term maintenance and renewal of forest plantations is technically difficult and is invariably costly. Theagroforest not only achieves a simple transfer of forest resources and structures. It also guarantees the renewability of these resources and structures, and of the related economic returns.

However, the agroforest strategy, empirically devised by swidden farmers all over the Indonesian archipelago, bears important weaknesses that could be solved by integrated research. The population of cultivated trees in the agroforest is usually genetically rather diverse, which affects the productivity of the crop. Though farmers do select the best producing individuals for reproduction, they cannot reliably capture genetic variation. Vegetative reproduction methods remain simple, and sanitary control in nurseries could be improved.

Technical research aimed at these weaknesses could reorient NTFP domestication for agroforestry in two ways. First would be the development of improved plant material specially designed for a complex, forest-like environment, rather than for conventional monocultural plantation conditions. Plant selection and breeding could be aimed at taking advantage of the 'forest' characteristic of the species for both ecological and economic benefits and adapting them to farmers technical, as well as energetic standards. The second direction for research would be to test high-yielding plant varieties-wild or improved-in agroforest conditions, as is being done by ICRAF in jungle rubber agroforests. This would expand the agroforest model to new areas and improved plant material into existing systems.

## Agroforest Domestication: A Socially Empowering Strategy for Farmers

The analysis of the agroforest domestication process should not be restricted to its technical or ecological aspects. While the transfer of the wild resources of nature to the cultivated lands of agriculture is an essential process-capturing variation in natural genetic characteristics, increasing population density, stimulating cross breeding, or escaping from natural competitors and pests-it must always integrate these with major economic and sociopolitical or policy implications. This could, in future NTFPs business, enable smallholders to do more than extract products from wild ecosystems, as they do now.

### From the forest to the fields: Who owns the resource?

By switching from the management of wild resources in traditional extractive systems to their adoption as new crops in farming systems, farmers often aim at maintaining or reestablishing their traditional authority over the forest resource base. This is obviously important when, because of overexploitation or deforestation, wild economic resources are vanishing from natural forests, or when commercial demand increases. But it might also be an important option for native farmers when politically induced dispossession threatens their livelihood. In most ideologies and political regimes of tropical countries, agriculture secures social and legal rights over land or natural resources for smallholders better than does forestry. Integrated forest management, as empirically conceived by indigenous forest tribes all over the planet, however sustainable or profitable it is, has never been seriously considered by the governing elites and their technical councils. This is reflected in the widespread lack of legal recognition of native rights and traditional property regimes concerning forest lands and resources in tropical countries.

To gain official support, native resource management systems have to evolve in a way that complies with the conventional models. Domestication and plantation are important steps in this process; transferring wild resources to cultivated lands is both a symbolic and a political act of appropriation. But beyond this conceptual aspect lies the legal context of appropriation. Most forest lands in the tropics, and the resources they contain, are under state control; they are 'public goods'. This usually prevents any evolution towards 'privatization' and facilitates tacit 'appropriation' of profitable forest resources, traditionally controlled by indigenous people, by those who are close to power. Private property-for either collective or individual owners-is more readily acknowledged and expected on agricultural land.

Thus, cutting the forest and planting trees might be enough to secure, if not property rights, at least the right to claim for such rights. In Indonesia, establishing agroforests has often been a major strategy for land and resource appropriation; establishing production structures and property rights that will be transmitted to further generations is an essential aspect in this particular domestication and cultivation process. For example, farmers in Sumatra initially planted damar-producing dipterocarp trees, in response to the depletion of wild damar trees and the need to establish a profitable forest-based economy. But as their relations with forest authorities deteriorated, the establishment of agroforests became a strategy for legal resource appropriation.

Thus, presently agroforests are also established as a claim against the closure of forest lands and resources to local communities. Through domestication and tree growing, farmers claim that they have purposefully restored and protected not only damar but the entire forest resource, in the middle of agricultural territory, upon which they hold a firmer control and rights. 'Global forest resource' appropriation is an essential component of the agroforest domestication strategy.

Specialized plantations might secure the appropriation of a given forest resource, but the agroforest strategy goes far beyond that: as it recreates forest structures, it allows the restoration of the landscape in a form that conforms better to the farmers' rights and interests.

In Indonesia, therefore, the relationship of local populations with forest resources is now more closely associated with one or more types of agroforest than with the natural forest. This allows them to maintain an economy and an associated lifestyle that remain in continuity with their forest culture, from which the agroforest directly evolved; but it places it firmly in an agricultural context. The agroforest, therefore, clearly opens the way for other novel models of improved resource management in forest lands throughout the tropics.

## Domestication and Cultivation: Knowledge and Capital

The choice for a particular domesticatior and cultivation strategy is as important as the transfer of forest resources to farmlands, in determining who holds the authority over a particular resource. Domestication and cultivation integrate technical knowledge and capital investments as well as technical, labour or energy inputs; all of these may be inaccessible to smallholder farmers. Colonial plantations have clearly demonstrated how domestication can adversely affect indigenous 'managers' of NTFPs. Will indigenous farmers be similarly spoiled when NTFP domesticates are made available by research institutes through markets or credit schemes? This seems likely when capital-intensive processes of crop establishment and maintenance lie far beyond smallholders' financial and technical capacities, and when the high productivity from plantations leads to a fall in prices of natural products and to the economic collapse of any business collecting those products. The domestication of NTFP through these sophisticated techniques and modern knowledge might only intensify the exclusion of smallholders from the management of forest resources.

In contrast, however, the domestication of NTFPs through the agroforest strategy will probably be better integrated into indigenous populations, as it relies on simple techniques, is based on local knowledge shared by every farmer, and does not imply high energy inputs. In the process of NTFP domestication, the evolution of rubber cultivation in Indonesia illustrates how plantation and agroforest development can have totally divergent effects on smallholders. Technical and financial constraints associated with the plantation model put rubber cultivation out of the reach of smallholders.

Thus, intensive rubber cultivation in colonial estates led to the exclusion of native tappers, both in Amazonia and Southeast Asia. However, when swidden cultivators in Sumatra and Borneo adopted the Para rubber plantation techniques to their production system by planting rubber trees in their swiddens, the trees grew with the fallow vegetation and soon evolved into a forest-like rubber garden. The trees were tapped if the prices appeared interesting, and they created an agroforest system that was much less demanding in labour and technical inputs than the current estate model. This production system soon became much more competitive than the estate plantations, and since 1945 it has gained the largest share in Indonesian rubber production. Through the rubber agroforests, former indigenous rubber collectors, evicted by the plantation owners, regained their place in the rubber trade and their share in the benefits of rubber development.

## The Forest Preference in NTFP Domestication: Some Economic Considerations

Domestication should not be disassociated from the global economic strategy of farmers. Future NTFPs plantations, whatever their model, will be part of lands claimed and developed through agricultural techniques. They will be integrated into agricultural territories and agricultural production systems. They will support local agricultural economy. The advantages of the agroforest model did not, until now, succeed in reliably capturing the trade of products from highly commercial species.

Short-term benefits are usually much higher when trees are grown in systems conforming to the plantation model. But so too are the risks and the energy consumption, with all its global ecological and economic consequences. The agroforest model, which by contrast emphasizes economic and ecological sustainability, should thus be evaluated in terms of long-term productivity, energy efficiency and economic security. It may then prove to be much more 'productive' and 'profitable' than the currently accepted models. Agroforests, in contrast to tree crop estates, allow the maintenance of numerous tree resources to grow together and to diversify the farmer's income.

If encompassed in the framework of agricultural strategies, agroforest development represents a process of forest conversion that does not go along with economic reductionism and that does not irreversibly close the economic potentialities formerly linked to the presence of natural forest. On the contrary, through the restoration of biodiversity in the agroforest, farmers maintain a whole range of economic choices for both the present and the future. Maintaining these options appears indispensable in view of the need for sustainable development. This multipurpose aspect must be kept in mind if systematic research on the domestication of NTFPs for agroforestry systems is to be carried out. An important aspect to consider in this respect is the potential for timber production. Timber will probably become a strategic commodity for farmers in the near future, with potential benefits that might be much higher than those provided by NTFPs. Many NTFP species also have good timber, but species domestication options, and the agroecosystem design, will unavoidably influence the capacity of the candidate forest species to produce quality timber. Investing in NTFP plantings will unavoidably lead farmers to some degree of specialization for a given product. However, opting for 'multipurposeness' in domestication and keeping in mind other potential forms of production, at both the species and the agroecosystem level, will help to avoid the irreversibility of future economic and ecological choices by smallholders.

## CONCLUSION, DOMESTICATION: TECHNIQUE OR STRATEGY?

As a domestication strategy based on forest resources, agroforest development represents an interesting alternative to the two common options devised for non-timber forest product management: harvesting from natural stocks or domestication for specialized plantations. Like specialized plantations, agroforests secure the conservation and multiplication of the planted-forest resources and increase the income-generating capacity of the forest. But they also ensure the restoration of a diverse forest ecosystem, as well as its integration into local agricultural production systems, while allowing local communities to maintain authority over its management. Plantations fail on several of these scores.

The NTFP domestication through agroforestry should not confine itself only to technical considerations. Besides field experiments, agroforestry research for NTFPs should focus on devising new strategies for better integration of forest resources into farmland, into rural economies and their related sociocultural, political and institutional systems. Until now, agroforestry research has never really dealt with long-lived trees, nor with true forest resources. Experimenting with long-lived trees obviously requires much more time than with fast-growing species, and new experimental designs will have to be found.

The integration of NTFP forest resources in better forms of land use will require not only new forms of experimentation but also new forms of conceptualization and implementation. Agroforests still lie on the margins of the conventional agroforestry research, despite the growing academic information about them. However, they touch the very heart of agroforestry, where forests and agriculture really meet and where forest structures and agricultural logic intersect. But agroforests are too close to forests for agriculturalists and too much embedded into farmers' activities for foresters.

This probably explains why conceptualization of agroforests is still denied by agriculture and forestry research. To deny their conceptualization is also to deny their existence and more importantly their future and its impact on future land use and forest resource management strategies. May the current research trends in domestication and commercialization of NTFPs give new opportunities for further development and integration of the agroforest concept.

# 3

# Industrial Hemp

## INTRODUCTION

Industrial hemp (*Cannabis sativa* L.) is also known as 'Indian hemp', 'cannabis' or 'hemp'. In Queensland, the *Drugs Misuse Act, 1986* refers to industrial hemp as 'industrial cannabis'. The plant has a long history and has been used for its bast (phloem) fibre in the stem, the multi-purpose fixed oil in the seeds (achenes), and an intoxicating resin secreted by epidermal glands. Items manufactured from it include food, textiles, paper, rope, fuel, oil, stockfeed, medicine, and spiritual and recreational products. It is thought that *C. sativa* was one of the first plants to be cultivated, and there is general agreement that the plant species originated in China, where the greatest diversity of germplasm is found. *Cannabis sativa* thrived in the manured soils around early settlements, which quickly led to its domestication. Early in the 20th century, industrial hemp was considered to be an important and beneficial crop throughout the Western World. However, as a result of the development of synthetics and its classification by the Western world as a drug in the 1920-30s, the production of industrial hemp was confined to India, Bangladesh and Eastern Europe.

Since the late 1980s there has been a resurgence of interest in fibre products, arguably driven by the green movement with a view to saving trees by growing renewable non-wood fibres in place of clearing forests for paper and building materials. This interest has gained momentum and credibility with Japan, for instance, setting a target date of 2010 to have 10% of its paper production sourced from non-wood fibre sources. Furthermore, the European Union has stipulated to its member countries that 95% of the components of each car produced by 2015 must be recyclable. The voluntary use by automotive manufacturers of natural fibres, including industrial hemp, suggests that demand for natural fibres will continue to increase.

Market segmentation for ethically produced goods, and growing support for biodegradable and natural products has led to a wide range of new industrial hemp products being developed. Although opportunities exist for industrial hemp, other fibre crops such as kenaf (*Hibiscus cannabinus* L.) may also play a significant role in meeting this demand. For a time, hemp was promoted as the ultimate crop from an economic and environmental perspective, requiring no pesticides, no fertiliser and no irrigation, as well as being a panacea for soil-borne disease.

Clearly, however, these claims do not reflect reality, and a far more pragmatic view of the crop is now emerging. Indeed, data from the FAO (2002) show that worldwide production of industrial hemp has steadily decreased since the 1970s. The legalisation of growing industrial hemp in some Australian states in recent years is recognition by government and the general community that industrial hemp may make a useful contribution to the economy as an alternative agricultural crop and that the crop can be grown under conditions that do not compromise law and order. Because there is no national policy for the crop, only *ad hoc* experimental trials and small-scale production have ensued. Licensing systems to allow for the commercial production of industrial hemp were first developed in Tasmania from 1991 and later in Victoria from 1997. Interest in the crop has declined in these states in recent years due to uncertainties about financial returns, marketing, and regulation. In Queensland, amendments to the *Drugs Misuse Act, 1986* were proclaimed in September 2002 to facilitate the commercial production of fibre and seed from industrial hemp crops.

Legislation to allow for the development of a commercial industrial hemp industry in Western Australia was passed in early 2004. Field trials have been conducted in NSW each year since 1995 and changes to the legislation in that state are now being sought with the view to commercialisation of the industry. It appears that Governments in South Australia and the Northern Territory are not proceeding to legislate for the commercial development of the industry at this stage. The objective of this chapter is to provide information to a wide audience on a broad range of topics relating to the industrial hemp industry.

Of particular interest are the issues that may affect the development of the industry in the Queensland context. As stated by Fletcher *et al*. (1995), industrial hemp has a large casual or devotional following because of its 'green' image and its products seem to be almost self-promoting. The main difficulty is determining which information is factual and which information is of a promotional nature.

## ABOUT THE PLANT

### Morphology and Anatomy

*C. sativa* is a tall, herbaceous annual plant with a deep tap root which grows to a height of up to 5 metres, depending on variety and growing conditions. The basic morphological features of the plant The stem is usually single and slender (4 to 20 mm diameter for mature plants) when grown at commercial crop densities. The stem tissues outside the vascular cambium are referred to as the bast (flexible inner phloem fibres of the bark) and contain the fibres useful for textiles The tissues inside the vascular cambium (inner woody core) contain the pith and the xylem vessels and are referred to as the hurd.

Industrial hemp is normally dioecious, meaning that the male and female flowers are borne on separate plants. Monoecious varieties, with separate male and female flowers on the same plant, have been bred and are widely cultivated throughout Europe. Monoecious varieties are reported to contain a proportion of dioecious plants, and also tend to have lower fibre yields than do the dioecious varieties. However, it has been claimed that monoecious varieties can produce a greater seed yield than can dioecious varieties.

## Difference Between Industrial Hemp and Marijuana — THC Concentration

Tetrahydrocannabinol (THC) is the classified psycho-active (mind-altering) ingredient in *C. sativa* that is produced in specialised glands (glandular trichomes). These glands are found primarily in the flowers surrounding the seeds, and, to a lesser extent, on the leaf surface of the plant. No such glands are produced on or in the seeds. The difference between marijuana and industrial hemp is that the THC concentration is significantly lower in industrial hemp than it is in marijuana. However, since seed is borne in the flowers that have a large number of glandular trichomes, traces of THC can cling to the seed hulls through the flower head's sticky resin. The concentration of THC varies according to environmental influences (such as oxygen, light, moisture, and temperature) and genetic factors.

It is generally accepted that industrial hemp plants are those *C. sativa* plants with a concentration of THC less than three per cent. The United States National Institute of Drug Abuse notes that most ordinary marijuana has an average THC concentration of three per cent. Much of the illicit cultivated *C. sativa* has a much higher THC concentration than three per cent. For example, sinsemal, also known as skunk, has a typical THC concentration range from 7.5 to 24%, although plants with a concentration higher than this range are also known to occur.

## Products and Markets

Traditionally, products made from industrial hemp fibre included rope and cordage, sailcloth, carpet backing, canvas, and apparel (such as the original Levi jeans made from hemp denim). Recent investigations have revealed that contemporary uses of industrial hemp may include reinforcing fibre for paper, fibre-reinforced plastics, polycomposites, fibreboards, geotextiles, textile fabrics (apparel and industrial), animal bedding, kitty litter, industrial absorbent products, and insulation. Many uses for the oil and seed have been developed or are under investigation, including animal stockfeed, soap, oil, paint and varnish, and cosmetics.

In some countries, (not Australia), the seed and oil from hemp plants are used in food products. In the European Union, cultivation of industrial hemp is more heavily weighted towards fibre than towards oilseed, with the production of about 27,000 tonnes of fibre versus only about 6,200 tonnes of seed in 1999. Conversely, the oilseed industry is the primary focus in Canada in recent years, with the breeding of new varieties and the development of improved technology for growing, harvesting and processing. Some of the products into which industrial hemp plants can be or are being made are listed below, and, where appropriate, mention is made of present markets or future market possibilities.

## TEXTILES

The history of textile production is rich and varied, with some 2,000 plant species having been processed into fibre at one time or other. Today, based on world production of fibre in 1999, 54.5% was synthetic (of which 60% was polyester), 42.9% was plant-based (of which 79% was cotton), and 2.6% was wool. In terms of plant fibre production other than cotton, flax is the only significant plant fibre crop and held 2.7% of the world plant fibre market. Only 0.3% of the world plant fibre production was derived from industrial hemp in 1999. For industrial hemp, the most desirable long fibres for textiles are found in the stem near the phloem tissue in the bast. Industrial hemp long fibre requires retting for preparation of high quality spinnable fibres for the production of fine textiles. Steam explosion is a technology that has been experimentally applied to industrial hemp. Using this technology, decorticated crude fibre is subjected to pressurised steam at high temperature to explode (separate) the fibres, resulting in hemp fibres that are thinner than those obtained from water retting. The refinement of equipment and new technologies as offering one possibility of making fine textile production from industrial hemp in developed countries, but noted that at present, China controls this market, and probably will remain dominant for the foreseeable future. Indeed, in the absence of the development of new technologies, considered that the concentration of spinning facilities and extraction technology in China, in addition to cheap labour, were major impediments for the production of industrial hemp fabrics outside of that country.

## PULP AND PAPER

The pulp and paper industry is currently based on wood fibre. Although industrial hemp fibre has been considered for use in pulp, it has only been used on an experimental basis. Since virgin wood pulp is required for added strength in the recycling of paper, the long fibres of industrial hemp could make paper produced from hemp fibre at least two times more recyclable than paper produced from wood fibre. However, various analyses have concluded that the use of industrial hemp for conventional paper pulp is not profitable. Indeed, because of a number of economic and structural issues, the use of fibre from industrial hemp for paper manufacture was considered to be unviable by Australian Newsprint Mills in partnership with the University of Tasmania between 1993 and 1996. The lower competitiveness of hemp as a source of paper pulp than that of wood sources reported an estimated price for hemp pulp of US$2,100 per tonne, compared with the price for bleached softwood pulp at a much lower US$800 per tonne.

Conversely, specialty pulp products made from industrial hemp, including cigarette paper, bank notes, technical filters, hygiene products, art paper, and tea bags, are believed to offer a highly stable, highly-priced niche market in Europe, where industrial hemp has an 87% market share of that sector. EIHA (2003) estimated that 17,000 tonnes of hemp fibre produced by affiliates of the European Industrial Hemp Association was used for pulp and paper applications in 2002.

### Plastic Composites for the Automobile and Other Manufacturing Sectors

Fibres may be introduced into plastics to improve their physical properties. Although manufactured fibres of glass and carbon are most commonly used. The amount of hemp used for automotive composites in Germany and Austria increased from zero tonnes in 1996 to an estimated 2,200 tonnes in 2002. Total use of natural fibres for automotive composites in these two countries increased over four-fold in the same time period (from 4,000 to 17,200 tonnes). It has been estimated that 5 to 10 kilograms of natural fibres can be used in the moulded portions of an average automobile (excluding upholstery).

### Building Construction Products

Industrial hemp fibres added to concrete increase tensile strength and reduce shrinkage and cracking. Fibre from industrial hemp is produced at a much higher cost than that from wood chips or straw from other crops. Given the greater strength of industrial hemp fibre than fibre from wood chips or straw, industrial hemp fibre may be more appropriately used in building materials requiring a high tensile strength.

## ANIMAL BEDDING/INDUSTRIAL ABSORBENT PRODUCT

Animal bedding products made from the hurd (inner woody core of the stem) of industrial hemp plants can absorb up to five times their weight in moisture, do not produce dust, and are easily composted. The high absorbency of hemp hurd has also led to its occasional use as an absorbent for oil and waste spill cleanup. That because hemp hurd is costly to produce (and animal bedding is a higher value use than industrial absorbent products), it is likely that animal bedding will remain the most important application of this product. An estimated 29,000 tonnes of hemp shives produced by members of the European Industrial Hemp Association were used to make animal bedding in 2002.

## GEOTEXTILES

Geotextiles include ground-retaining, biodegradable matting designed to prevent soil erosion, especially to stabilise new plantings while they develop root systems along steep highway banks to prevent soil slippage, or ground covers designed to reduce weeds in planting beds. The economic viability of using industrial hemp for geotextile applications is yet to be determined. However the relatively high cost structure would suggest that it may be difficult for industrial hemp to penetrate this market.

## STOCKFEED

Expression of oil from the seed of industrial hemp plants leaves behind a protein-rich, oil-poor seed cake, also referred to as 'seed meal'. This seed meal has proven to be an excellent source of nutrition for animals and does not contain THC, which is present in the leaves and flowering heads of industrial hemp plants. The seed meal from 5,000 tonnes of hemp seed produced by members of the European Industrial Hemp Association was used in 2002 for animal feed.

Under the *Stock Regulation, 1988* (Queensland) there are restrictions imposed on the feeding to livestock of certain cannabis because THC is a fat-soluble compound and is known to occur in the milk of animals consuming feed that contains THC. It follows that it is probable that this compound may also be in the fat of such animals.

The following is a summary of cannabis that a person must NOT feed to stock or NOT allow stock to gain access:

- any cannabis plant that still has leaves, flowers or seed attached;
- 'failed' industrial cannabis crops left in a growing paddock unharvested;
- viable cannabis seed capable of producing cannabis plants (seed that has not been denatured).

The following is a summary of cannabis that a person may feed to livestock or allow livestock to gain access:

- industrial cannabis stems or ground industrial cannabis stems after harvesting or treating industrial cannabis plants.( plants containing no more than 1% THC in the leaves and flowering heads). In other words the plant after all leaves, flowering heads and seeds have been removed;
- denatured seed from industrial cannabis plants grown by a licensed grower;
- the oil extracted from processed industrial cannabis; and
- seed meal ground from denatured industrial cannabis seed.

If you are a licensed grower of industrial cannabis plants and livestock animals are being reared on either your own or your neighbour's property then reasonable measures must be taken to deny these animals access to your industrial cannabis crop. If necessary fencing should be carried out to make sure livestock cannot accidentally stray into industrial cannabis crops.

## FOOD IN THE HUMAN DIET

Industrial hemp is not an approved food product in Australia. On 24 May 2002, the Australia New Zealand Food Standards Council (FSC) (Food Regulation Secretariat 2002) rejected an application for the inclusion of industrial hemp seed and oil in novel food. The FSC is comprised of health ministers from each Australian state and the heads of the Australian and New Zealand governments. The decision was taken despite the recommendation by Food Standards Australia New Zealand (FSANZ, formerly ANZFA) that the total prohibition on the use of *Cannabis* spp. in food be removed. The rationale for the decision by the Ministers of FSC was that there were law enforcement issues, particularly from a policing perspective, where it was perceived that there would be difficulties in distinguishing between high THC and low THC products. The Ministers of FSC also believed that the use of industrial hemp in food may send a confused message to consumers about the acceptability and safety of *C. sativa*.

Consistent with the decision by FSC, Section 5 of the Queensland *Drugs Misuse Act 1986* requires that a product manufactured from industrial cannabis must be in a form that stops it from being consumed by humans (as well as in a form that stops it from being smoked or administered). Despite the restriction in Australia, about half of the world market for oil extracted from industrial hemp seed is currently used for human food and food supplements (de Guzman 2001). In North America, many of the products from the seed are incorporated into food preparations such as snack bars, bread, pretzels, biscuits, yoghurts, pancakes, porridge, ice cream, pasta, pizza, salad dressings, mayonnaise and beverages (Small and Marcus 2002). Such foods currently have a niche market, based particularly on natural food and specialty food outlets.

### Personal Care Products

In the 1990s, European firms introduced lines of hemp oil-based personal care products, including soaps, shampoos, bubble baths, and perfumes. The hemp oil is now marketed throughout the world in a range of body care products, including creams, lotions, moisturisers, and lip balms. In Germany, laundry detergent manufactured entirely from hemp oil has been marketed. Hemp-based cosmetics and personal care products account for about one-half of the world market for hemp oil (de Guszman 2001). Of the approximate one billion US dollars in gross sales that is reported annually by The Body Shop, about 4% of sales in 2000 were hemp products.

## Production Trends

The information relating to the size of the international market for industrial hemp is not complete and in some cases is not reliable. Official data (FAO 2002) of worldwide area harvested and total production show that the industry has declined substantially over the past 30 years. Sources other than FAO indicate that areas of industrial hemp harvested are usually greater than those published by the FAO. This discrepancy illustrates the problems that confound any attempt to characterise the international market for those crops that are currently of lesser importance in western industrialised nations. Reliable and current information for such crops is often difficult to find. Nevertheless, the declining trend in world-wide production is unequivocal, regardless of the method used to determine it.

China is the largest producing country of hemp fibre, with over half of total world production extracted from plants grown in that country. Other significant producing countries include India, the Russian Federation, and North Korea.

## Issues

Several authors have noted many of the issues confronting the industrial hemp industry in Australia. In general, these issues include the need for more research, financial/economic considerations, a lack of processing infrastructure, a requirement to better assess market demand, and better access to markets.

## Research and Funding

Research to develop varieties with low photoperiod sensitivity, reduced THC concentration, and a high yield of fine fibre is seen as a critical priority for the industrial hemp industry in Australia. Furthermore, research into optimum sowing times and densities, nutrient application rates, photoperiod response, susceptibility to pests and disease, the impacts of wind and extremes in soil moisture, the ability of the crop to act as a weed suppressant, yield and quality of fibre produced, crop rotations and the opportunity cost of planting compared with other crops have also been seen as important. Some of these issues are being addressed in research programmes conducted in Queensland and elsewhere. However, the resources of funding bodies to support such research programmes is seen by some as inadequate and a major impediment to the development of the industry.

### Financial/Economic/Price

Numerous issues impact on the economic viability of the industrial hemp industry in Australia and its competitiveness on world markets. A major concern is competition from the European Union and China, the major world producers of the crop. The European Union has heavily subsidised production, and labour costs in China are only a fraction of those in Australia. Furthermore, products from other bast fibre species such as flax, kenaf, or jute suitable for the paper pulp, textile or woven fabric industries can compete with those from industrial hemp without the need for monitoring and inspection services.

Retted jute is imported into Australia for carpet backing for a cost of AU15 cents per kilogram (approximate figures derived from Customs import data;). Woven fabrics (such as for hessian bags) cost AU50 cents per kilogram, and yarn that is processed further costs AU$1.30 per kilogram. Producers of commodities with small markets are generally at the mercy of small fluctuations in price or quantity. Small increases in the world production of hemp fibre caused export prices to fall by half to a world average of US35 cents per pound in 1996 (Vantrese 1998). Variation in the price of hemp fibre has also been considerable, with the hemp yarn price varying from US$1.01 to US$3.31 per kilogram from 1993-99.

### Processing Infrastructure

Many of the end uses of industrial hemp require separation of the bast and hurd fibres or the extraction of oil from seed. Separation of fibres or extraction of oil are primary stage processes that would require milling facilities to be located close to production areas to minimise the cost of transportation of the bulk material. Such facilities require large capital investment and are not yet available at the required scale.

## Market Demand

Industrial hemp products have found a place in niche markets in the developed world. There is speculation of more opportunities in the future, with some industry proponents estimating the international market for bast fibre to increase from 100,000 tonnes in 1999 to over 20 million tonnes by 2050. However, the prediction of market size has been fraught with difficulty in the past, with a dramatic over-estimation of market potential by Canada in the late 1990's, with around 14,000 hectares grown. The degree to which industrial hemp will meet market demand will depend on demonstrable proof that its products are of equal or superior quality to its competitors, at equal or reduced costs in adequate and consistently available amounts.

## Access to Markets

It has been estimated that approximately half of the world market for industrial hemp oil is currently for human food and food supplements. The comparatively higher profitability of producing seed and oil for this market than of producing plant products for other markets has been suggested. Current prohibition in Australia and New Zealand of industrial hemp seed and oil in novel food denies access to this potentially lucrative market.

## Gross Margins/likely Profitability/Financial Risk

Industrial hemp plants have been widely reported to produce high yields and to have a strong resistance to pest incursion. Such claims, if substantiated, augur positively for keeping costs down and profitability up. However, most gross margin analyses for industrial hemp include little or no consideration of the cost of necessary monitoring and inspection by regulatory staff, which, in Queensland, can exceed $1,000 per crop annually.

Furthermore, processing requires substantial capital investment that is usually not factored into analyses of industry profitability. One of the major determinants of profitability is market price, or more important to the grower, the farm-gate price. Hitherto, such prices are not well defined, largely because markets are still being developed. Nevertheless, gross margin assessments have been made, and the reader needs to be cognisant of the estimates that are used, the assumptions made, and the data that are or are not included. Much of the contemporary published information on the profitability of the industrial hemp industry is somewhat dated, since many of the references were published in the mid to late 1990's, a time prior to commercial production in Queensland.

The dearth of more recent data, however, does not necessarily mean the conclusions made in those studies are no longer relevant. Several authors have highlighted the dearth of key information on the potential returns from industrial hemp production. For example, Fitzgerald (1995) expressed frustration with this fact, and stated that data on market prices, market size, and production costs were especially difficult to obtain. He stated that the industry in Australia had come about by a combination of ignorance and uncertainty, since the market demand was difficult to ascertain, more agronomic assessment was required, primary processing technologies were expensive and not suitable for anything other than a niche industry in Australia, current market prices were 'spot prices' which could not be used to determine economic rates of return for an expanded industry in the future, and any resurgence of industrial hemp production in Australia might alter significantly the world pricing structure and therefore have a substantial impact on returns. The gross margin for the production of decorticated material for the dry geotextile market for feedstock for the paper pulp market was to range from $60/hectare for a yield of 1 tonne dry stem/hectare to $452/hectare for a yield of 12 tonne dry stem/hectare.

The analysis assumed a fibre value of $300/ tonne, and did not include the cost of irrigation or inspection fees. For a straw price of $130/ tonne and input costs (excluding irrigation and inspection fees) of $360/ hectare, RIRDC estimated a break-even yield of 2.8 tonne dry straw/ hectare. It was clear that industrial hemp did not provide a gross margin comparable with many current crops within potential growing regions. It was assumed that market penetration would be difficult for all paper pulp types, with the exception of those suited to specialty papers. For both the paper pulp and textile markets, the production of high quality bast fibre was an expensive and difficult process and would not be feasible unless by-product markets were secured. A study by Australian Hemp Resource and Manufacture (1997), now called Ecofibre Industries Limited (EIL), provided a far more favourable assessment of the potential returns from growing industrial hemp. Using variable costs for producing dryland sorghum in Central Queensland as well as information from trials conducted in southern Australia and overseas, these authors estimated gross margins for industrial hemp for northern NSW and Queensland of $1,164 to $3,354/ hectare.

It is concluded that textile products offered the highest potential for value-adding to industrial hemp, but warned that suitable processing facilities for hemp textile production were not available in Australia. Because the domestic market for natural fibres was small and mostly supplied by the cotton and wool industries, both of which export the majority of production, it was believed that a viable industrial hemp industry would also need to export the majority of production. Furthermore, RIRDC called into question the cost of processing industrial hemp compared with competing fibres, since the yield of high value fibre from harvested stems was relatively small at less than 1% by weight.

A SWOT (strengths, weaknesses, opportunities, threats) analysis of the commercial prospects for the development of an industrial hemp industry in Australia. Although the analysis was based on information that is now somewhat dated, he concluded that there was considerable interest and public support for the development of a hemp industry in Australia. However, he considered that the demand for fibre was almost entirely from potential producers or conservationists, rather than from those involved in the processing or marketing. In a review of the potential of industrial hemp products for industrial and commercial applications, it is concluded that seed and oil crops would produce the most economically viable products. Furthermore, she stated that the residual stems and fibres of the harvested crop could be utilised for a range of products, including paper, building materials, or biomass fuel.

The current prohibition of industrial hemp seed and oil in novel food by the FSC, despite a recommendation by FSANZ that the total prohibition on the use of low THC C. *sativa* in food be removed, denies the industry access to this potentially lucrative market. Assessments of the profitability of growing industrial hemp give widely varying accounts of likely financial performance. Opportunities in specific niche industries such as specialist paper fabric and board production have been identified. However, a lack of information and a view by some that the industry is high risk are issues that need to be considered carefully in the business plan of any potential participant.

## Legislative Frameworks/Industry Development in Australia

Enthusiasm for an industrial hemp industry has grown in Australia in recent years. Legislative frameworks developed by some State Governments have given the industry the opportunity for broad-scale commercial production and/or for the development of research programmes. Although there has been small-scale experimental production of industrial hemp under a permit system in most States in the last several years, no nation-wide policy has been developed for the crop. This lack of a coordinated national approach has led to a large number of *ad hoc* trials being conducted for a range of purposes, including variety assessment, machinery development, and the determination of better agronomic practices.

A synopsis of legislative and industry development activities that have been taken place to support the development of an industrial hemp industry in each Australian state and the Northern Territory is provided below.

## The *Drugs Misuse Act 1986* of Queensland defines *C. sativa* as a dangerous drug

For the purpose of assessing the potential of *C. sativa* plants with a low THC concentration for commercial fibre production, the Act was amended in early 1998. These amendments allowed for controlled field trials and plant breeding research to be conducted for a period of three years. The trial period was subsequently extended for twelve months by the one extension permitted under section 43Y of the Act. The extended period expired on 18 December 2002.

On 8 August 2002, the *Drugs Misuse Amendment Bill* was passed by Parliament, which allowed for the commercialisation of industrial hemp fibre and grain. Proclamation of the amendments to both the *Drugs Misuse Act 1986* and the *Drugs Misuse Regulation 1987* occurred on 27 September 2002.

The legislation now allows for the research, production, processing, marketing, and trade of processed industrial cannabis fibre and seed products in Queensland, with the exception of those products that could be smoked, administered or consumed.

Prior to the amendments to the Act, possession or supply of commercial industrial hemp products that were available in the market place (e.g. hemp shirts, hemp hand cream etc) was technically in breach of the law, despite the fact that the THC concentration may have been minimal or non-existent. The amendments ensured that sections 5 (trafficking in dangerous drugs), section 6 (supplying dangerous drugs), section 8 (producing dangerous drugs) or section 9 (possessing dangerous drugs) of the Act do not apply to a manufactured product. A 'manufactured product' is defined as a product made from, or partly from, processed cannabis with a THC concentration of not more that 0.1% and that is in a form that cannot be inhaled, administered or consumed. This last requirement prevents a person from trying to sell products that mimic illegal products such as 'low hemp' cigarettes.

## An Outline of Legislative Requirements

The amendments to the *Drugs Misuse Act 1986* are underpinned by the requirement to hold either a researcher licence and/or a grower licence or to be an authorised person. Only licensed or authorised persons are able to deal with C. *sativa* without committing an offence under the Act in relation to trafficking in, supplying, producing, publishing or possessing instructions for producing and possessing C. *sativa.* The licensing system differentiates between legal and illegal use of C. *sativa.* Exemptions from offences under the Act only operate while licensees perform activities in accordance with the Act and the conditions of the licences. Industrial hemp inspectors are appointed under the Act (division 11) to monitor the compliance of licensees with their licence. Inspectors have wide powers of entry (with or without consent) and general powers. There are three categories of licences, each one permitting the use of C. *sativa* with different concentrations of THC, as follows:

1. A *category 1 researcher licence,* among other things, enables a person to possess for research purposes industrial cannabis plants and seeds and class A and class B research cannabis plants and seed. A 'class A research cannabis plant' has been defined to mean a cannabis plant that has a THC concentration in its leaves and flowering heads of 3% or more. Class A research cannabis seed has been defined to mean seed harvested from a Class A research cannabis plant or seed that, if grown, will produce a class A research cannabis plant. A 'class B research cannabis plant' has been defined to mean a cannabis plant with a THC concentration in its leaves and flowering heads of more than 1% but less than 3%. Class B research cannabis seed' has been defined to mean seed harvested from a class B research cannabis plant or seed that, if grown, will produce a class B research cannabis plant. A category 1 researcher licence enables the holder to source new strains of C. *sativa* plants from the wild into their plant breeding programmes. Strict security provisions apply to this category of licence, and plant breeders need to demonstrate appropriate educational or other qualifications and experience to participate in this activity.

2. A *category 2 researcher licence,* among other things, enables a person to possess for research purposes industrial cannabis

plants and seed and Class B research cannabis plants and seed. Plant breeders need to demonstrate appropriate educational or other qualifications and experience to participate in this activity.

3. A *grower licence*, among other things, enables a person to possess industrial cannabis plants and seed and to produce industrial cannabis plants from certified cannabis seed. An 'industrial cannabis plant' has been defined to mean a cannabis plant with a THC concentration in its leaves and flowering heads of not more than 1%. Industrial cannabis seed has been defined to mean cannabis seed harvested from an industrial cannabis plant or certified cannabis seed. 'Certified cannabis seed' has been defined to mean seed certified, in the way prescribed under a regulation that will produce cannabis plants with a THC concentration in their leaves and flowering heads of not more than 0.5%. However, licensed growers are authorised to possess industrial cannabis plants and seed with a THC concentration of up to 1% to allow for circumstances in which elevated THC concentrations that may result from climatic or environmental changes. Without some tolerance levels growers would be exposed to criminal prosecution.

## Eligibility to Hold a Licence

To be eligible to hold a grower licence, an applicant (if a corporation, its executive officers), must not have been convicted of a serious offence in the preceding 10 years and not be affected by bankruptcy action. To be eligible to hold a researcher (category 1 or 2) licence, an applicant (if a corporation, its executive officers) must not have been convicted of a serious offence in the preceding 10 years and have the necessary educational or other qualifications and experience to engage in plant breeding or other research. For a corporation applying for a researcher licence, a person employed by the corporation to carry out plant breeding under the licence who is not an executive officer meets this criterion. A serious offence is defined in the legislation. The Chief Executive of DPI&F also has the power to determine the suitability of an applicant having regard to their character, honesty and integrity and the character of their close associates (close associate is defined in the legislation), their criminal history and whether they are capable of satisfactorily performing the activities of a licensee. Applicants for a licence are required to undergo a criminal history check by the Queensland Police Service.

### Provision for Authorised Persons

In addition to the licensing requirements of growers and researchers, the *Drugs Misuse Act, 1986* protects from drug trafficking certain other persons (authorised persons) who need to possess, supply or transport industrial hemp, ancillary to licensed growers and researchers. Authorised persons include denaturers, seed suppliers, analysts, carriers, inspectors, manufacturers and family members and employees of licensed holders. Certain conditions are attached to these persons and these conditions are outlined in the *Drugs Misuse Regulation, 1987*. The conditions imposed on authorised persons are less stringent than those imposed on licence holders because of the lower risks involved.

### Costs Associated with Licences, Inspection, and Monitoring

As set by regulation, an application fee must accompany an application for a licence. Applicants are also required to pay the cost of criminal history checks for themselves as well as their close associates. For corporations, criminal history checks are required for each executive officer. A fee to enable the Queensland Police Service (QPS) to conduct these criminal history checks is also required. However licence applicants send their criminal history check fees with their licence application fee direct to DPI&F and not QPS. After three years, a licensee has the option of renewing their licence for a further three-year period. A renewal fee will apply. For details of the licence application fee, licence renewal fee or the fees for criminal history checks that are currently in force please contact the DPI&F Business Information Centre.

Participation in the industrial hemp industry is based on a user-pays principle. The legislation requires the licence-holder to pay the reasonable costs of compliance monitoring activities performed under the licence, including inspection and plant sampling fees and any laboratory analysis necessary to determine the concentration of THC in the leaves and flowering heads of cannabis plants in the possession of the licence holder. Licence-holders should expect to have their crops analysed for THC concentration. Where more than one variety of industrial hemp is grown, each variety is analysed for THC concentration. The cost of inspections is based on the time taken to conduct the inspection, as well as the travelling time of the inspector. It would therefore be prudent for growers to make provision for an amount of $1,000 as a minimum in their annual commercial hemp production budgets to cover these inspection and analytical costs. Where repeat inspections are necessary due to non-compliant behaviour, the licensee is required to pay the full cost of these additional checks.

## Victoria

Full commercialisation of the industrial hemp industry in Victoria has been possible since 1997, when the *Drugs, Poisons and Controlled Substances Act, 1981* was amended to allow the production of low-THC cannabis for non-therapeutic use. The Act, administered by Victoria's Minister for Health and the Department of Human Services, allows for the possession, cultivation and selling of industrial hemp by authorised persons who meet strict criteria and undergo a police records check. Three-year-authorisations are issued by the Department of Primary Industries Victoria, and the Secretary of that department must be notified of changes in ownership of management of the business of the person. Authorisations are also subject to various terms, conditions, limitations and restrictions. Significant fees and charges apply. Most of the permits that have been issued since 1997 were for less than one hectare, although one authorisation was approved for 30 hectares. According to Nowland (2002), the total area of industrial hemp planted in Victoria was less than 100 hectares per year. Various trials have been conducted for example, field assessments conducted at five sites in 1996-97 showed a weight of stems (dried in the field) of cultivar Futura 77 of 5.3 to 12 tonnes/hectare. Interest in the industrial hemp industry in Victoria has declined significantly in recent years, with only two authorisations current for 2003-04.

### New South Wales

Field trials for research purposes have been permitted in NSW since September 1995 under the *Drug Misuse and Trafficking Act, 1985* that is administered by NSW Health. The allowable upper limit for THC in trial crops is set at 0.3%, and the Act does not allow for commercial production of the crop. Trials have been conducted at a range of sites, and fibre and seed production have been assessed. The overall objective of the trial programme is to assess the yield potential of low-THC crops in a range of environments within NSW and, where possible, to have the stem fibre or seed produced by the crop evaluated for paper, textile, food and other products.

The NSW Agriculture administers and processes applications for the field trials and supervises trial sites. Applicants meet all costs associated with the trials. About 40 trials have produced an average yield of 5 tonnes dry stems/hectare, which is not considered to be economically viable (Nowland 2002). However, trials conducted during 2000-01 produced yields of 12 tonnes dry stems/hectare (Spurway and Trounce 2003). The highest yields occurred in the central west of the state using a centre-pivot overhead sprinkler system. However, most of NSW is deemed to have insufficient summer rain for a dry-land crop to be an option. Ten trials were approved for 2003/04, bringing the total number of trials conducted since 1995 to 68.

Hitherto, only industrial hemp trials have been conducted in NSW, and the committee for approvals generally limits these trials to 5 hectares or less, although some growers make requests for a larger area than this amount. With recent reports of yields of 12 tonnes/hectare, the committee for approvals is moving closer to seeking changes to the legislation to allow for the commercialisation of the industry. It is believed that despite intense competition from subsidised overseas production, there is some chance of competing on the world stage (pers. comm. Bob Trounce, NSW Agriculture).

### Tasmania

The Tasmanian Hemp Company based in southern Tasmania, was first licensed in 1991-92 to cultivate industrial hemp for commercial research purposes. Australian Hemp Research and Manufacture (AHRM), based in Queensland, commenced trials in Tasmania in the mid-1990's and was later joined by a newly formed group called the Tasmanian Hemp Growers' Cooperative. AHRM has now changed its name to Ecofibre Industries Limited (EIL). Licences are granted by the Department of Health and Human Services in cooperation with Tasmania Police and the Department of Justice and Industrial Relations, in accordance with the provisions of the *Poisons Act, 1971*, which is presently under review. There is provision for the licensing of processors, however, no large commercial processing facility exists in Tasmania to date. Cottage industry size manufacturing of specialty stationery products does occur. Areas sown to industrial hemp between 1997 and 2002 have fluctuated from 2 to 40 hectares. For 2003/04, only 0.2 hectares was sown (pers. comm. Joe Horak, DPIWE Tasmania). Yields in Tasmania have been approximately 5 to 6 tonnes fibre/hectare or 800 kilograms seed/hectare.

Confidence in the Tasmanian industry has fluctuated dramatically over the years. This fluctuation appears to be dependent upon financial, marketing and regulatory variability, particularly due to recent decisions regarding hemp for food (pers. comm. Joe Horak, DPIWE Tasmania). DPIWE continues to process inquiries regarding the production of industrial hemp, although it is expected that a lack of processing facilities will continue to stymie development of the industry in the short term.

## Western Australia

From 1996-97 until 1999/2000, industrial hemp trials were conducted in Western Australia through exemptions granted under the *Poisons Act, 1964* and the *Misuse of Drugs Act, 1981*. There have not been any growers of industrial hemp since the Department of Agriculture Western Australia ceased its trials in 1999-2000. Legislation to allow for the development of a commercial industrial hemp industry in Western Australia, the *Industrial Hemp Bill, 2003*, was introduced into the Western Australian Parliament in November 2003 and was passed in early 2004. Under the new laws, licences to cultivate, harvest or process industrial hemp are issued by a Registrar, appointed for that purpose. Background checks are conducted on all applicants to determine their character and any criminal associations. The licensing conditions determine the location of crops, ensure security measures are put in place to restrict access to seeds and plants, and articulate the conditions for harvesting and processing. The new legislation allows the police and specially appointed inspectors to enter and inspect properties, examine seed, plants or crops and remove them for testing.

## South Australia

Three trials to assess the growth of industrial hemp were licenced by the South Australian Government in the mid-1990's. Irrigation trials conducted in the south east of the State in late spring demonstrated sufficient yield to promise a commercial potential, with subsequent testing required. Attempts to grow industrial hemp as a dryland winter sown crop on Yorke Peninsula and the Lower North failed. The trial programme has recently been wound down due to low trial yields. While the current legislation in SA is restricted to allowing only research trials, reported that the South Australian Government sees little point in changing the law to make it easier to grow commercial crops until real commercial prospects can be demonstrated.

## Northern Territory

At the present time, the Northern Territory Government is reported to have no intention of licensing any industrial hemp trials as it sees short day lengths as a major problem during the growing season. It was reported that kenaf is considered to be a better option than industrial hemp and a commercial proposition was encouraged from a South East Asian company to grow this rival crop for use in the paper industry. However, due to a lack of interested growers, the kenaf project is now on hold.

## Yield

### *Fibre*

Yields of dry stems harvested in trials at Biloela, Queensland, have typically ranged from 8 to 11 tonnes/hectare. The dry weight of stems of industrial hemp plants grown during the 2003 season in trials in the Mackay district ranged from 7.5 to 9.1 tonnes/hectare, depending on variety and locality. It is hoped that higher yields than these will be attained through breeding and biotechnology. Gaining access to germplasm from around the world will be essential to maintain a competitive edge and enhance new product development. An average crop of industrial hemp grown in France to produce 6 to 8 tonnes stem (16% moisture)/hectare. This yield would equate to a bast fibre yield of 2.1 to 2.8 tonnes/hectare, assuming an average 35% to 40% bast in each stem.

Fourty trials of industrial hemp in NSW produced an average yield of 5 tonnes dry stems/hectare, although a yield of 12 tonnes dry stems/hectare was reported for irrigated plants grown in trials in the central west of that State. In 2001 and 2002, an average yield of dry hemp stalks of 5.4 to 6.2 tonnes/hectare was produced in the European Union by companies aligned with the European Industrial Hemp Association from a cultivated area of 10,400 hectares.

### Oil Seed

Ideally, the yield of industrial hemp seed per unit area should be based on the air-dry weight of the seed, which is usually about 12% moisture. Exaggeration of up to 50% of the yield of industrial hemp is sometimes the result of measuring the fresh weight of harvested seed rather than recording the weight of dry seed. Yield of 1 tonne/hectare is usually considered good in the European Union, whereas in Canada, where considerable effort has been given to develop the oilseed industry, yields of up to 1.5 tonne/hectare have been achieved. In 1999, the average yield of seed from industrial hemp crops grown in Canada was 0.9 tonne/hectare.

## AGRONOMY — FARMING THE CROP

It may be that to achieve a viable gross margin, the production of industrial hemp in Australia will need to be undertaken on a broadacre basis. If this assumption is correct, the size of industrial hemp farming operations would need to be comparable with other broadacre crops such as cotton, wheat or barley. Furthermore, planting, cultivation and harvesting equipment would need to be of sufficient capacity to handle large-scale operations.

Information on growing industrial hemp is more focussed on fibre crops than on oilseed crops. The following information on the cultivation of industrial hemp is not sufficient to address all of the practical difficulties that may emerge and should be treated as a general guide only.

## Soil

Industrial hemp plants grow well on a fertile, neutral to slightly alkaline, well-drained clay loam or silt loam soil (Reichert 1994). The requirement for a well-drained site is necessary as industrial hemp plants are particularly sensitive to wet, flooded, or waterlogged soil. This susceptibility has been noted from plantings conducted in Tasmania and in Queensland.

## Temperature, Photoperiod

For most commercial varieties, industrial hemp is a short day plant, meaning that flowering occurs when the daily period of light is shorter than some critical length (e.g. 13 hours). High temperatures also accelerate flowering. Some varieties of industrial hemp that have been bred in high-latitude countries with comparatively long photoperiods have critical day lengths which are close to those of the maximum day length experienced in Queensland (summer solstice). Growing such varieties in Queensland results in premature flowering and a low fibre yield due to a shift of photosynthetic assimilates away from vegetative growth to reproductive growth.

## Varieties

Although there are over 2,000 accessions of *C. sativa* known to exist, only a small number of varieties have low concentrations of THC. Finding varieties with low THC concentration and suited to Queensland conditions is an important step in developing a successful industry. Whereas varieties currently grown in the European Union may be suited to the growing latitudes of Tasmania or Victoria, they would not necessarily be adapted to

Queensland. It is possible that varieties from Pakistan, India, China, and South America may be more appropriate than European varieties for Queensland, although such lines are renowned for having an unstable THC concentration and little is known of their commercial potential in terms of the proportion and quality of fibre in stems when using full mechanisation.

Breeding programmes for industrial hemp varieties suited to Queensland are being conducted. Recent trials have shown that these programmes are delivering lines upon which a viable Queensland industry can be based.

## Sowing

A well-prepared seedbed that is fine, level and firm is best for uniform germination. Analysis of the soil prior to sowing is recommended, the seeding rate for industrial hemp has been widely researched in a number of centres and recommendations include 40 to 150 kilograms seed/hectare for fibre production and 1 to 24 kilograms seed/hectare for seed production. The fibre strains are typically sown at a minimum rate of 250 seeds per square metre (approximately 45 kilograms/hectare), and up to three times that density is sometimes recommended. Seeding rates for grain production vary widely, from 10 to 45 kilograms/hectare. The use of a roller at sowing may assist germination by facilitating good contact between the seed and the soil. Seeds are normally sown in rows spaced from 70 to 200 mm apart.

The optima metre have been found to have little effect on stem dry matter yield. For Queensland conditions, a population density of 90 to 200 plants per square metre is considered desirable for fibre crops.

## Irrigation

Until germination has occurred (usually 3 days after sowing), it is recommen specially during the first six weeks of its growth. Without rain, a crop may require 2 to 6 ML of irrigation water/hectare.

## Pests and Disease

Industrial hemp has a reputation for being resistant to pests and disease, although the degree of resistance has been greatly exaggerated, with the crop playing host to several insects and fungal pathogens. Grey mould, caused by the fungus *Botrytis cinerea*, is one of the most significant diseases associated with industrial hemp, and there are nearly 300 pests worldwide, the most serious of which are the European core borer (*Ostrinia nubilalis*) and the hemp borer (*Grapholita delineana*).

Green vegetable bug (*Nezara viridula* Linnaeus), heliothis moth (*Helicoverpa* sp), and monolepta beetle (*Monolepta australis* Jacoby) have been found in crops of industrial hemp grown for fibre in Queensland, but appear not to have had a significant impact on yield, possibly because the rapid growth of the crop tends to minimise the effects of pest damage. However, damage to the terminal buds of plants, particularly from heliothis moth, may require strategic pest control intervention. Monolepta beetle was reported to have defoliated a large proportion of the industrial hemp plants grown in a trial conducted in the Hunter Valley of NSW. Under optimal growing conditions, such as a well-prepared seed bed, adequate soil moisture, rapid germination, and a high plane of nutrition, canopy closure normally occurs 5 to 7 weeks after sowing. This competitive ability of industrial hemp plants against weeds usually obviates the need for herbicide application during the life of the crop. However, sound agronomic practices for weed management prior to planting need to be followed to reduce the competitive effect from the weed population early in the life of the crop.

## Use of Pesticides

There are no pesticides registered or available for use under off-label permit in industrial hemp in Australia. Therefore, no pesticide may be legally applied to the crop. Should the use of a particular pesticide be required, it would be necessary for the industry to make application for an off-label permit to the Australian Pesticides and Veterinary Medicines Authority (APVMA). The Biosecurity business group of DPI&F can assist industry with submitting the permit application. Information on applying for an off-label permit may be obtained from the APVMA website by selecting the 'permits and minor uses' link.

## Harvesting

For fibre production, industrial hemp plants are normally cut in the early flowering stage or while pollen is being shed, well before seeds are set. The stem, bark and fibre yield of industrial hemp plants reached their maximum at the time of flowering of the male plants, a stage of development that was called 'technical maturity'. However, it is showed that a later harvest time at the beginning of seed maturity led to easier decortication without any effect on the tensile strength of the bast. A stem diameter of 4 to 7 mm is considered ideal for fibre production, and most hemp billets are chopped to a length of 40 to 60 cm. Consideration of mechanical harvesting technology suitable for industrial hemp has only recently begun to take place in developed countries such as Australia where the high cost of labour means that it is necessary to harvest mechanically to effectively compete with countries with relatively low labour costs.

## Retting

The traditional and still major first step in fibre extraction is to ret (rot) away the softer parts of the plant, by exposing the cut stems to microbial decay in the field (dew retting) or by submerging them in water (water retting). The result of retting is the sloughing off of the outer parts of the stem and to loosen the hurd from the phloem fibres.

For dew retting, harvested industrial hemp is placed into windrows and left to desiccate until fungal organisms can complete the retting process in the stems. The length of time for the retting process will depend entirely on the weather, but typically requires one to 2 weeks. Windrows would normally be turned once or twice, and if not turned, the stems close to the ground remain green while the top ones are retted and turn brown. Determining when the stalks have become sufficiently retted requires experience - the fibres should turn golden or greyish in colour, and should separate easily from the interior hurd. Stalks should have less than 15% moisture when baled, and should be allowed to dry to about 10% in storage out of the weather. Water retting has been largely abandoned in countries where labour is expensive or where environmental regulations exist. Water retting, typically by soaking the stalks in ditches, can lead to a high level of pollution.

Nevertheless, most hemp fibre currently used in textiles is still water retted in China and Hungary. The development of improved processes such as by the use of micro-organisms or enzymes in water, or by industrial ultrasonic retting could augment or replace traditional methodology. However, such techniques are still being researched or have been carried out in semi-commercial situations for other crops such as flax. It is reported that pilot projects have shown the energy, water, and chemical cost of enzyme or ultrasonic breakdown processes to be approximately $1.95/ kilogram ($1950/ tonne) of treated fibre, without including the costs of infrastructure, finance or labour. On this basis, therefore, only dew retting would be considered economically viable for Australian conditions at this time.

## Agronomic Issues for Queensland

Poor plant stand is one of the issues that has confronted some farmers in growing industrial hemp in Queensland. It is believed that poor seed-bed preparation, incorrect sowing depth, and inappropriate use of irrigation water following sowing may have contributed to poor crop establishment. Poor germination and plant establishment has also been reported in trials conducted in NSW, with reasons varying from low seed viability, high temperature, or use of unsuitable European varieties.

Although generally one crop of hemp can be grown each year, the climate in Queensland can lend itself to two crops per year, provided the first planting occurs during September to October. However, decreasing day-length after the summer solstice (21 to 22 December) may cause premature flowering and stunted plant growth leading to a poor yield of fibre.

## CONCLUSIONS

The strong world demand for renewable and recyclable fibre products and the decline and demise of some established agricultural industries in recent years has created an interest in industrial hemp and an opportunity for diversification into other crops. The legalisation of growing industrial hemp in some Australian states in recent years is recognition by government and the general community that industrial hemp may be a viable alternative crop and that it can be grown under conditions that do not compromise law and order. Although it has been argued by some that industrial hemp is low risk from an agronomic perspective, that it has the requirement for low chemical input and that it makes an excellent rotation crop, the question is whether or not the industry will be profitable for growers and other participants.

As for most new crops, the incidence of pests and disease is lower than that of established crops that provide an available food source and the opportunity to develop resistance to various control strategies. However, not often disclosed is the fact that should they be needed, pesticides are not registered or available for use under off-label permit in industrial hemp crops in Australia, and applying these formulations could lead to prosecution. By submitting an application to the APVMA for an off-label permit, access to use a specific pesticide can be obtained on a temporary basis. The occurrence of failed industrial hemp crops in both Queensland (poor plant stand) and NSW (low yield) may go some way to emphasise the need for employing good crop husbandry and management practices. Poor agronomic practice was almost certainly a contributing factor in Queensland, and poor adaptation of varieties originating from regions in the northern hemisphere with long daylengths may be a possible explanation for the poor outcomes in both States.

Breeding programmes by EIL and Crop Tech Research Pty Ltd may address the latter of these issues in the future. Although the majority of industrial hemp in the world is grown or processed using low mechanisation and high labour inputs at minimal cost, in Australia the industry most likely will need to be broadacre, mechanised and capital intensive to achieve viable gross margins. Publicly available assessments of the profitability of the industrial hemp industry give widely varying accounts of likely financial performance.

Opportunities in specific niche industries such as specialist paper fabric and board production have been identified. One view is that if the industry is to develop and prosper in Queensland, it will be through the development of markets for innovative new products rather than through the more traditional uses such as paper and cloth. However, a lack of information and a view by some that the industry is high risk are issues that need to be considered carefully in the business plan of any potential participant. From a broader economic perspective, the success of the industry in Australia depends on competition from other countries, such as the European Union that has subsidised production, and China that has low labour costs. Moreover, other bast fibre crops such as flax and kenaf will also compete with industrial hemp, without the need for monitoring and inspection fees and charges. Some researchers have concluded that the seed and oil from industrial hemp crops could produce the most economically viable products, while utilising the stems and fibres as a residual agricultural product.

Despite the fact that about half of the world market for industrial hemp oil is currently for human food and food supplements, the prohibition in Australia and New Zealand of industrial hemp seed and oil in novel food denies access to this potentially lucrative market. Particularly for fibre and perhaps for oil, processing infrastructure needs to be located within close proximity to growing areas to minimise the cost of transport of the bulk material. Such facilities, which require a large capital outlay, are not yet available at the required scale in Australia. Speculation of a future burgeoning market for bast fibre products is a positive sign for the industrial hemp industry in Queensland.

However, the degree to which industrial hemp can share in this assumed growth in market demand will depend on its products being comparable or of equal quality to its competitors, at equal or reduced costs in adequate and consistently available quantities. It can be argued that the industry is in a 'Catch 22' - until the market is well defined and the magnitude of demand is known, a lack of investment in capital intensive processing infrastructure will be an impediment.

Conversely, however, until there is adequate, consistent, and cost competitive supply, demand will not be crystallised. Industrial hemp plants can now be grown commercially in Queensland, Tasmania and Victoria. Legislation to allow for the development of a commercial industry in Western Australia was passed in early 2004, and changes to the legislation in NSW are being sought with the view to commercialise the industry. Although industrial hemp has been grown commercially in Tasmania and in Victoria since the early and mid-1990s, respectively, the industry is currently in decline in those States. Further research and development initiatives need to include plant breeding to improve THC stability in heat tolerant varieties, to increase stalk and fibre yield, and to widen the adaptability of the varieties that are available. Additionally, identifying suitable growing locations and streamlining farming and processing systems would also be useful initiatives for development of the industry.

# 4

# Economic Importance of Plants

## INTRODUCTION

Plants are extremely important in the lives of people throughout the world. People depend upon plants to satisfy such basic human needs as food, clothing, shelter, and health care. These needs are growing rapidly because of a growing world population, increasing incomes, and urbanization. Plants provide food directly, of course, and also feed livestock that is then consumed itself. In addition, plants provide the raw materials for many types of pharmaceuticals, as well as tobacco, coffee, alcohol, and other drugs.

The fibre industry depends heavily on the products of cotton, and the lumber products industry relies on wood from a wide variety of trees (wood fuel is used primarily in rural areas). Approximately 2.5 billion people in the world still rely on subsistence farming to satisfy their basic needs, while the rest are tied into increasingly complex production and distribution systems to provide food, fiber, fuel, and other plant-derived commodities. The capability of plants to satisfy these growing needs is not a new concern. The Reverend Thomas Malthus (1766-1834) in his *Essay on the Principle of Population* in 1798 argued that population growth would exceed nature's ability to provide subsistence. According to the U.S. Census Bureau, the world population was about one billion in 1800, doubled to two billion in 1930, doubled again to four billion in 1975, and reached six billion people in 2000.

World population is expected to be nine billion by the year 2050. The challenge to satisfy human needs and wants still exists. Income has also been increasing rapidly throughout most of the world at the same time. U.S. census estimates are that the gross national product reached $27,000 per person in 1997 and is expected to reach $69,000 in 2050 assuming a 1.8 percent annual rate of growth. Income per person in many countries of Asia, Latin America, and Africa has increased more rapidly, but continues to be less than in other areas such as Western Europe and the United States. As income grows, plants become more valuable because people want to buy more and higher-quality products to satisfy basic needs.

Increasing urbanization leads to an increase in demand for marketing services as populations relocate from rural areas to urban areas. According to the Census Bureau, the U.S. population, for example, changed from 60 percent rural in 1900 to less than 25 percent rural in 2000. This urbanization demands more marketing services to assemble, sort, transport, store, and package large quantities of foods from production centers to consumption centers.

## VALUE OF PLANTS

According to the United Nations Food and Agriculture Organisation, the estimated export value of major plant commodities traded in world markets for 1998 was: rice ($9.9 billion dollars), maize ($9.1 billion), wheat ($15.1 billion), soybeans ($9 billion), coffee greens and roast ($13.7 billion), sugar ($5.9 billion), tobacco ($24.1 billion), cigarettes ($15.4 billion), lint cotton ($8.2 billion), forest products ($123 billion), and forest pulp for paper ($13 billion). Markets, a place where people buy and sell goods and services, determine the economic value of plants. The value depends on the expected uses and benefits provided. The economic value of plants is measured by their prices in a market economy. Demand and supply determine the price. In most countries, markets operate freely with little direct government interference in trading. In centrally planned economies such as China, however, the government frequently controls market operations, and buys and sells through government companies.

In planned economies, governments may set prices administratively at levels that do not indicate true economic value to consumers and producers.

As world economies become more open and market-oriented through trade agreements such as those that come from the World Trade Organisation, the value of plants will likely become more equal among countries. Two main types of markets set the value of plants: cash markets and future markets. The most common type is a cash market. Cash markets are very popular places throughout the world where buyers and sellers meet to exchange money for goods and services. Demand and supply in the cash market set the price at which buyers will exchange money with sellers for immediate possession of goods. In the simplest case, producers take goods to the market for immediate sale, and consumers arrive with cash to buy goods for immediate possession.

In more complex cases, producers sell goods to one or more other buyers who in turn sell the goods to consumers. Cash markets operate daily, weekly, or for other intervals all over the world. Consumers and producers trade in thousands of cash markets operating in the world today. These local cash markets in rural areas are linked to larger regional trading centers that in turn are linked to cash markets in the larger cities. Cash markets operate for all the major plant products. Futures markets, a second major market to set the economic value of plants, operate very differently from cash markets. In cash markets buyers and sellers trade the physical good, whereas in futures markets buyers and sellers trade a futures contract.

Futures contracts are standardized written documents calling for future delivery of a good at a particular time and place in the month of expiration. Futures markets attempt to discover the best value today for a good tomorrow based upon expected demand and supply in some future time period. Futures markets have become increasingly popular around the world. Important futures exchanges include the Chicago Board of Trade for grains and oilseeds; the Chicago Mercantile Exchange for livestock, dairy products, and lumber; and the New York exchanges for cocoa, coffee, cotton, orange juice, and sugar. Futures contracts are traded on exchanges in Great Britain, France, Japan, Australia, Singapore, and Canada. In addition, Brazil, China, Mexico, Italy, and Spain (to name a few) have futures exchanges. It is interesting to note that futures markets do not trade contracts in fruits and vegetables and other highly perishable products. Futures trading is not possible for highly perishable products because of the difficulty of long-term storage.

## MARKETING SYSTEMS FOR PLANTS

The marketing system for most plants can be viewed as an hourglass shape that concentrates production from many farms into large quantities and fewer firms for processing and handling, followed by a distribution into smaller quantities and more firms for sale to many consumers. Marketing systems add value as the plants progress from the farmer to the consumer. The added value takes the form of marketing services that transform a raw commodity into a finished product for consumer use. Depending on the commodity, these services include cleaning, sorting, grading, packaging, storing, transporting, handling, processing, and financing until goods are sold to the consumer. Farmers usually sell their goods at harvest time in local markets to buyers who may come from large urban or smaller regional trading centers, or farmers sell to agents of those buyers. The buyers assume the risks of ownership until they are able to sell the goods to consumers at a later time.

The ownership risks include providing many valuable marketing services to assure that products will be available in the right quality, in the right place, at the right time, in the appropriate amount, and for a reasonable price. The difference between the value paid by consumers for plants and the value received by producers is the marketing margin, which is the amount charged by the businesses for the services provided. For example, if the consumer pays one dollar for a product in the supermarket, and the producer receives forty cents, then the marketing margin is sixty per cents.

Higher incomes and growing populations mean that consumers will demand more marketing services that increase convenience and reduce preparation time, such as slicing, freezing, packaging, and ready for microwaving. In addition, as per capita income increases, the composition of demand for food changes to increased consumption of higher-value products. These changes typically mean increased consumption of products such as fruits and vegetables, meat, dairy, and processed products, and decreased consumption of staples such as potatoes, cassava, and rice.

More marketing services are required for high-valued products. As consumers demand more marketing services, the marketing margin will increase, causing the farmers' share of the consumer food dollar to decline. In many countries, the farmers' share of consumer expenditures (about 40 to 50 percent) is already declining, as marketing margins increase. John Abbott in *Agricultural and Food Marketing in Developing Countries* indicated that margins also vary by country for the same commodity due to differences in income, geography, infrastructure, and marketing systems. The farmers' share of consumer food expenditures has declined steadily through time in the United States to about 21 percent in 1993; ranging from 25 percent for food consumed at home to 15 percent for away-from-home consumption. This declining farmers' share can be expected to continue as income increases.

A declining farmers' share does not mean that the marketing system is inefficient or that farming is unprofitable. Technical progress that increases productivity generally will result in declining real prices per unit of output. Farmers can increase their share of the consumer food expenditures by adding value to what they sell. Some examples of added value are direct sales to consumers at farmers' markets, roadside markets, and farmer-owned marketing and processing cooperatives. Paul Eck in *The American Cranberry* described Ocean Spray cranberry juice as a most successful story of farmers adding value to cranberries. Cranberry growers formed a cooperative to process and market Ocean Spray cranberry juice more profitably, a product that has great brand identification with consumers.

# 5

# The Importance of Plant Life

## INTRODUCTION

Interest in exotic and native aquatic plants is quickly growing as water gardening climbs to the top of gardening activities throughout the country. Statistics show water gardening quickly growing at an annual rate of nearly 20% since 1997, with no sign of dwindling. Increasing demand means there's more aquatic plants to be found in today's market. Water gardener's now have their choice of hundreds of aquatic plants to choose from; water lilies, lotus, hyacinth, cattails, rushes and hundreds more. Water lilies alone now come as day-bloomers, night-bloomers, hardy and tropical, with flower colors ranging from white, pink, red, yellow, purple, blue and orange in hues from pastels to bright vibrant colors. Many of the tropical lilies are very fragrant bringing them to top place in elite water garden choices. For aquatic plant lovers, attractiveness isn't all that's at chance when it comes to landscaping a water garden.

## ALL PLANTS PLAY A VITAL ROLE IN WATER GARDENING!

Aquatic plants are an important addition to the water garden for balance as well as beauty. A blend of aquatic plants help keep the pond ecologically balanced, they provide needed shade, cool and clean water that is vital to a healthy pond. For the average home water garden, a good rule is to have at least 50% of the water garden's surface covered in plants, a variety of submersed plants such as water lilies, golden poppy, floating heart, etc. in which the leaves come to the surface creating shade, and floating plants such as hyacinth, water lettuce, duckweed, etc. are only a part of the choices available to water gardeners to create this balance. Aquatic plants help keep water clear by playing an important part in biological filtration, this is especially important with the submerged plants such as hornwort, anacaris and dwarf sagittaria species. Through various means, aquatic plants utilize excess nutrients and waste products such as nitrogen, nitrates and ammonia out of water that would otherwise cause algae growth, or high levels of toxins. Marginal and submerged plants play different roles in providing shade, food and protection for fish and other aquatic wildlife, while also providing breeding and nesting places for fish, frogs, insects and birds.

Following is a list of the major types of plants for use in and around a water garden:

- *Hardy Lilies* - Hardy lilies are reliably perennial from USDA zone 3 througzone 11. Hardy lilies typically bloom from May through September. When cold weather comes, the foliage dies and sinks to the bottom. New leaves will begin to arise from the submerged rhizomes the following spring.
- *Hardy Marginal Plants* - Marginal plants grow along the perimeter of ponds, lakes, wetlands, and streams. Marginals soften the hard look of rocky edges and create a smooth transition from the water to the planting areas surrounding the pond. Marginals help shade a pond and can be part of the natural filtration of a bog area. Most marginal plants like one to eight inches of water. These plants are recommended to be placed on shelves within the pond area in containers as most all pond plants are aggressive in nature.
- *Tropical Lilies* - Tropical lilies produce fragrant and vibrant colorful blooms. The flowers, usually carried above the water surface on strong stems, come in brilliant whites, yellows, pinks, reds, and lilacs. Tropical lilies are only hardy in USDA zones 10 and 11. In colder zones, tropicals are most often

treated as annuals, though they can be brought inside and over-wintered with proper care. Tropical lilies are either day bloomers or night bloomers. Day bloomers flower in the early morning and close in the late afternoon/early evening. Nighbloomers open in the late afternoon/early evening and stay open until early next morning, and are especially popular with the working crowd as they tend to spend their time in the gardens in the late afternoons or early evenings.

- *Tropical Marginal Plants* - Tropical marginal plants originate from subtropical and tropical regions and are not perennial in cooler climates. A common practice is to leave these tropicals inside their pots and stack small cobblestones around the outside to disguise the pot.
- *Lotus* - The lotus is the King of the water plants, noted for its magnificent flowers. These exquisite flowers have circular leaves and fragrant flowers. Even their seedpods are interesting, and often are used in dried flower arrangements. This plant should be kept in a pot, as it is highly aggressive. It must be submerged to the deepest part of the pond in winter to avoid death from freezing.
- *Floating Plants* - These plants, which float on the water's surface with their roots dangling below, provide shade for pond water and also absorb excess nutrients there thus making summer algae control easier. Most are tropical, but some are hardy. In colder climates, tropical floaters should be treated as annuals to be replanted each year or over-wintered indoors. In ponds with skimmers, avoid planting where floaters can drift into and clog the skimmer.
- *Water Lilies* - The water lily is the Queen of the water garden and the most popular water gardening plant. Lilies provide a lovely floating carpet of leaves and flowers across the surface of a pond. They also provide fish with a shady canopy from the hot summer sun while starving algae of the sunlight needed for growth. There are winter hardy and tropical varieties.

- *Hardy Submerged Plants* - Submerged plants are often the most overlooked plants in the water garden. Because they are under water, they typically don't grab the onlooker's attention like the lilies and marginal plants do. Yet, these plants are important allies in creating a well-balanced water feature. Also known as oxygenating plants, submerged plants help reduce algae by directly competing for food sources. Think of them as nutrient sponges. They provide food, protection and coverage for baby fish and other aquatic wildlife. They can be planted by simply pushing a bundle of plants right into the gravel.

## PLANT INFORMATION

The use of cattails in your pond are numerous for wildlife as well as humans. Red-wing blackbirds, coots, marsh wrens, and other shore birds nest among cattails, giving them a safe place. Cattails can be soaked in oil to be used as torches. Cattail stems are edible either raw or boiled. Native American's used the juice from the stem to remedy an aching tooth by the rubbing juice onto the gums. The fluffy 'catkins' were also used by Native Americans for their softness and absorbency for diaper material for infants.

The cattail is one of the most important and common wild foods, with a variety of uses at different times of the year. Whatever you call it, a stand of cattails is as close as you get to finding a wild supermarket. By late spring, the light green leaves reach nearly nine feet tall, forming a sheath where they tightly embrace the stalk's. The leaves hide the new flower head until it nears maturity. Peel them back to reveal it. The plant is so primitive dating back to the time of the dinosaurs that male and female flowers are separate on the stiff, two-parted flower head: the pollen-producing male is always on top, while the seed-bearing female is forever relegated to the bottom. Clearly, this species evolved long before the Sexual Revolution. (Biological speaking, this arrangement is effective because the male part withers away when its job is done, whereas the female part must remain connected to the rest of the plant until the seeds have matured and dispersed.)

Cattails grow in dense stands. Like most colonial plants, they arise from rhizomes thick stems, growing in the mud, usually connecting all the stalks. A cattail stand is like a branching shrub lying on its side under the mud, with only the leaves and blossoms visible. The two most widespread species in the United States are the common cattail (*Typha latifolia*), which is larger and bears more food, and the narrow-leaf cattail (*Typha augustifolia*).

*Caution:* Young cattail shoots resemble non-poisonous calamus (*Acorus calamus*), and poisonous daffodil (*Amaryllidaceae*) and iris (Iris species) shoots, which have similar leaves. If a stand is still topped by last year's cottony seed heads, you know you have the right plant. In spring, the cattail shoot has an odorless, tender, white, inner core that tastes sweet, mild, and pleasant far cry from the bitter poisonous plants, or the spicy, fragrant calamus. None of the look-alikes grows more than a few feet tall, so by mid-spring, the much larger cattail becomes unmistakable, even for beginners.

Cattails grow in marshes, swamps, ditches, and stagnant water fresh or slightly brackish worldwide. Finding them is a sure sign of water.The cattails every part has uses. It's easy to harvest, very tasty, and highly nutritious. It was a major staple for the American Indians, who found it in such great supply, they didn't need to cultivate it. The settlers missed out when they ignored this great food and destroyed its habitats, instead of cultivating it. Before the flower forms, the shoots prized as "Cossacks asparagus" in Russia are fantastic. You can peel and eat them well into the summer.

They're like a combination of tender zucchini and cucumbers, adding a refreshing texture and flavor to salads. Mix them with pungent mustard greens to balance their mildness. Added to soup towards the end of cooking, they retain a refreshing crunchiness. They're superb in stir-fry dishes, more than suitable for sandwiches, and excellent in virtually any context.

Harvest cattail shoots after some dry weather, when the ground is solid, in the least muddy locations. Select the largest shoots that haven't begun to flower, and use both hands to separate the outer leaves from the core, all the way to the base of the plant. Now grab the inner core with both hands, as close to the base as possible, and pull it out. Peel and discard the outermost layers of leaves from the top down, until you reach the edible part, which is soft enough to pinch through with your thumbnail (the rule-of-thumb). There are more layers to discard toward the top, so you must do more peeling there. Cut off completely tough upper parts with a pocket knife or garden shears in the field, so you have less to carry.

*Note:* Collecting shoots will cover your hands with a sticky, mucilaginous jelly. Scrape it off the plant into a plastic bag, and use it to impart a slight okra-like thickening effect to soups. The shoot provide beta carotene, niacin, riboflavin, thiamin, potassium, phosphorus, and vitamin C. The proportions of food to waste varies with the size of the shoot. You'll get the best yield just before the flowers begin to develop. A few huge, late-spring stalks provide enough delicious food for a meal. Some stalks grow tall, and become inedibly fibrous with developing flowers by late spring, although just before the summer solstice, you can often gather tender shoots, immature flower heads, and pollen at the same time.

You can clip off and eat the male portions of the immature, green, flower head. Steam or simmer it for ten minutes. It tastes vaguely like its distant relative, corn, and there's even a central cob-like core. Because it's dry, serve it with a topping of sauce, seasoned oil, or butter. It's easier to remove the flesh from the woody core, if desired, after steaming. This adds a rich, filling element to any dish, and it's one of the best wild vegetarian sources of protein, unsaturated fat, and calories. It also provides beta-carotene and minerals.

When the male flowers ripen, just before the summer solstice, they produce considerable quantities of golden pollen. People pay outrageous prices in health stores for tiny capsules of the bee pollen a source of minerals, enzymes, protein, and energy. Cattail pollen beats the commercial variety in flavour, energy content, freshness, nutrition, and price. To collect the pollen in its short season, wait for a few calm days, so your harvest isn't scattered by wind. Bend the flower heads into a large paper bag and shake it gently. Keep the bag's opening as narrow as possible, so the pollen won't blow away. Sift out the trash, and use the pollen as golden flour in baking breads, muffins, pancakes, or waffles. It doesn't rise, and it's time-consuming to collect in quantity, so generally it is mixed with at least three times as much whole-grain flour. You can also eat the pollen raw, sprinkled on yogurt, fruit shakes, oatmeal, and salads.

During fall, winter, and early spring, the cattail rhizomes store food. Digging up the thick, matted rhizomes from the muck, especially in cold weather, is not easy. Peel off the rhizomes outer layers, still imbued with mud, then work the starch from the fibers with your fingers, in a large bowl of water. The water became cloudy with the starch. Wait an hour to let the starch settle, and pour off the water, getting enough sweet, tasty starch to thicken a small pot of soup hardly worth the effort.

An alternate method is to tear apart the washed rhizomes and let them dry, pound the fibers to free the stach, and sift. This yields as about as much starch as the previous method. I've also tried chewing on the fibers inside the cleaned rhizomes and swallowing the starch, which is very tasty. However, the digging and cleaning is so much work, I'd have to be starving in the winter to bother. Furthermore there are reports that eating the starch of some species raw may cause vomiting. The buds of the following year's shoots, attached to the rhizomes, are also edible. Although they make a tasty cooked vegetable, they are a bit too small to be worth digging up and cleaning, although their size may also vary.

Collecting the flower heads and pollen doesn't harm the plant, because cattails spread locally by their rhizomes the seeds are for establishing new colonies, and each flower head makes thousands of these. Collecting a small fraction of the shoots also does no damage, since the colony continually regenerates new shoots. Since nobody wants to sink into the mud, people normally collect at the periphery of the stand. Of course, if the stand is small, it's already struggling to survive adverse conditions. Finding a larger stand elsewhere will increase your harvest, and give the embattled plants a chance.

The Indians also cattails medicinally: They applied the jelly from between the young leaves to wounds, sores, boils, carbuncles, external inflammations, and boils, to soothe pain. Besides its medicinal uses, the dried leaves were also twisted into dolls and toy animals for children, much like corn-husk dolls found today. Cattail leaves can be used to thatch roofs, weave beautiful baskets, as seating for the backs of chairs, and to make mats. Archeologists have excavated cattail mats over 10,000 years old from Nevada caves. No longer edible once the pollen is gone, the brown flower heads make good "punks," supporting a slowly-burning flame, with a smoke that drives insects away. The fluffy, white seeds were once used for stuffing blankets, pillows and toys. The Indians put them inside moccasins and around cradles, for additional warmth. Cattails and their associated microorganisms improve water and soil quality. They render organic pollution harmless, and fix atmospheric nitrogen, bringing it back into the food chain. They've even been planted along the Nile river to reduce soil salinity.

# 6

# Root Vegetable

## INTRODUCTION

Root vegetables are plant roots used as vegetables. Other underground plants are often, erroneously, called root vegetables. Root vegetables include both true roots such as tuberous roots and taproots, but exclude non-roots such as tubers, rhizomes, corms, and bulbs. Several types contain both taproot and hypocotyl tissue, and it may be difficult to distinguish the two. Contrary to popular belief, vegetables are not true root vegetables unless their names are appended by "ruta" (i.e. rutabaga). Regardless of anatomical type, root vegetables are generally storage organs, enlarged to store energy in the form of carbohydrates. They differ in the concentration and the balance between sugars, starches, and other types of carbohydrate.

Of particular economic importance are those with a high carbohydrate concentration in the form of starch. These starchy root vegetables are important staple foods, particularly in tropical regions. They overshadow the cereals throughout much of West Africa, Central Africa, and Oceania, where they are used directly or mashed to make *foufou* or *poi*. Some Jains are opposed to eating root vegetables for ethical reasons.

## LIST OF UNDERGROUND VEGETABLES BY ANATOMICAL TYPE

### True Root

- Taproot (some types may incorporate substantial hypocotyl tissue)

- *Apium graveolens* (celeriac)
- *Arctium* spp. (burdock or gobo)
- *Arracacia xanthorrhiza* (arracacha)
- *Beta vulgaris* (beet and mangelwurzel)
- *Brassica* spp (rutabaga and turnip)
- *Bunium persicum* (black cumin)
- *Daucus carota* (carrot)
- *Lepidium meyenii* (maca)
- *Pachyrhizus* spp (jicama and ahipa)
- *Pastinaca sativa* (parsnip)
- *Petroselinum* spp. (parsley root)
- *Raphanus sativus* (daikon and radish)
- *Scorzonera hispanica* (black salsify)
- *Sium sisarum* (skirret)
- *Tragopogon* spp. (salsify)

**Tuberous Root**

*Conopodium majus* (pignut or earthnut) *Ipomoea batatas* (sweet potato) *Manihot esculenta* (cassava or yucca or manioc) *Mirabilis extensa* (mauka or chago) *Psoralea esculenta* (breadroot, tipsin, or prairie turnip) *Smallanthus sonchifolius* (yacón

**Modified Stem**

- Corm
- *Amorphophallus konjac* (konjac)
- *Colocasia esculenta* (taro)
- *Eleocharis dulcis* (Chinese water chestnut)
- *Ensete* sp. (enset)
- *Sagittaria* sp. (arrowhead or wapatoo)
- *Xanthosoma* sp. (malanga, cocoyam, tannia, and other names)
- Rhizome

- *Zingiber officinale* (ginger, galangal)
- *Curcuma longa* (turmeric)
- *Panax ginseng* (ginseng)
- *Arthropodium* spp. (rengarenga, vanilla lily, and others)
- *Canna* sp. (canna)
- *Cordyline fruticosa* (ti)
- *Maranta arundinacea* (arrowroot)
- *Nelumbo nucifera* (lotus root)
- *Typha* sp. (cattail or bulrush)
- Tuber
- *Apios americana* (hog potato or groundnut)
- *Cyperus esculentus* (tigernut or chufa)
- *Dioscorea* sp. (yams, ube)
- *Hemerocallis* sp. (daylily)
- *Helianthus tuberosus* (Jerusalem artichoke or sunchoke)
- *Lathyrus tuberosus* (earthnut pea)
- *Oxalis tuberosa* (oca or New Zealand yam)
- *Solanum tuberosum* (potato)
- *Plectranthus edulis* and *esculentus*. (kembili, dazo, and others)
- *Stachys affinis* (Chinese artichoke or crosne)
- *Tropaeolum tuberosum* (mashua or añu)
- *Ullucus tuberosus* (ulluco)

## VEGETABLE

The term "vegetable" generally means the edible parts of plants. The definition of the word is traditional rather than scientific, however, and therefore the usage of the word is somewhat arbitrary and subjective, as it is determined by individual cultural customs of food selection and food preparation. Generally speaking, a herbaceous plant or plant part which is regularly eaten as unsweetened or salted food by humans is considered to be a vegetable. Mushrooms belong to the biological kingdom Fungi, not the plant kingdom, and yet they are also generally considered to be vegetables, at least in the retail industry.

Nuts, seeds, grains, herbs, spices and culinary fruits are usually *not* considered to be vegetables, even though all of them are edible parts of plants. In general, vegetables are those plant parts that are regarded as being suitable to be part of savory or salted dishes, rather than sweet dishes. However there are many exceptions, such as the pumpkin, which can be eaten as a vegetable in a savory dish, but which can also be sweetened and served in a pie as a dessert. Some vegetables, such as carrots, bell peppers and celery, are eaten either raw or cooked; while others are eaten only when cooked. For many different kinds of vegetables, please see the list of vegetables

**Is it a fruit or a vegetable?**

The word "vegetable", in its modern usage, is strictly a culinary term, and is *not* a botanical or scientific term. The word "fruit" on the other hand can be a culinary term, or it can be a botanical term, and these two usages are quite different. Botanically speaking, fruits are fleshy reproductive organs of plants, the ripened ovaries containing one or many seeds. Thus, many botanical fruits are not edible at all, and some are actually extremely poisonous.

In a culinary sense however, the word "fruit" is applied only to those botanical fruits or similar plant parts which are both edible and palatable, and which in addition are considered suitable to be a sweet or dessert food, such as strawberries, peaches, plums, etc. In contrast to this, a number of edible botanical fruits, including the tomato, the eggplant, and the bell pepper, are not considered to be sweet or dessert foods, are not routinely used with sugar, but instead are almost always used as part of a savory dish, and are salted. This is the reason that they are labeled as "vegetables".

Thus in a scientific context, a plant part may correctly be termed a "fruit", even though it is used in cooking or food preparation as a vegetable. The question *"The tomato: is it a fruit, or is it a vegetable?"* found its way into the United States Supreme Court in 1893. The court ruled unanimously in *Nix v. Hedden* that a tomato is correctly identified as, and thus taxed as, a vegetable, for the purposes of the 1883 Tariff Act on imported produce. The court did acknowledge however that botanically speaking, a tomato is a fruit.

Some vegetables defined as different parts of plants:

- *Flower bud*: broccoli, cauliflower, globe artichokes
- *Seeds*: Corn also known as maize
- *Leaves*: kale, collard greens, spinach, beet greens, turnip greens, endive
- *Leaf sheaths*: leeks
- *Buds*: Brussels sprouts
- *Stems of leaves*: celery, rhubarb (sometimes loosely called a "fruit", because sweet pies are made from it)
- *Stem of a plant when it is still a young shoot*: asparagus
- *Underground stem of a plant, also known as a tuber*: potatoes, Jerusalem artichokes, sweet potato (often incorrectly called a yam in the USA), also the true yam.
- *Whole immature plants*: bean sprouts
- *Roots*: carrots, parsnips, beets, radishes, turnips
- *Bulbs*: onions, garlic, shallots
- *Fruits in the botanical sense, but used as vegetables*: tomatoes, cucumbers, squash, pumpkins, capsicums (bell peppers and hot peppers), eggplant, tomatillos, christophene, okra, including tree fruits eaten as vegetables, such as breadfruit and avocado, and also the following:
  - *Legumes*: peas, beans
  - *Whole unripe seed pods of legumes*: green beans, snap peas

## ETYMOLOGY

"Vegetable" comes from the Latin *vegetabilis* (animated) and from *vegetare* (enliven), which is derived from *vegetus* (active), in reference to the process of a plant growing. This in turn derives from the Proto-Indo-European base **weg-* or **wog-*, which is also the source of the English *wake*, meaning "become (or stay) alert".

The word was first recorded in print in English in the 14th century. The meaning of "plant grown for food" was not established until the 18th century.

The word is still sometimes used as an archaic literary term for any plant, as in *vegetable matter, vegetable kingdom.*

### In the Diet

Vegetables are eaten in a variety of ways, as part of main meals and as snacks. The nutritional content of vegetables varies considerably, though generally they contain a small proportion of protein and fat, and a relatively high proportion of vitamins, provitamins, dietary minerals, fiber and carbohydrates. Many vegetables also contain phytochemicals which may have antioxidant, antibacterial, antifungal, antiviral and anticarcinogenic properties.

### Colour Pigments

Vegetables (and some fruit) for sale on a street in Guntur, India.

The green color of leafy vegetables is due to the presence of the green pigment chlorophyll. Chlorophyll is affected by pH and changes to olive green in acid conditions, and bright green in alkaline conditions. Some of the acids are released in steam during cooking, particularly if cooked without a cover.

The yellow/orange colors of fruits and vegetables are due to the presence of carotenoids, which are also affected by normal cooking processes or changes in pH.

The red/blue coloring of some fruits and vegetables (e.g. blackberries and red cabbage) are due to anthocyanins, which are sensitive to changes in pH. When pH is neutral, the pigments are purple, when acidic, red, and when alkaline, blue. These pigments are very water soluble.

Many root and non-root vegetables that grow underground can be stored through winter in a root cellar or other similarly cool, dark and dry place to prevent mold, greening and sprouting. Care should be taken in understanding the properties and vulnerabilities of the particular roots to be stored. These vegetables can last through to early spring and be nearly as nutritious as when fresh.

During storage, leafy vegetables lose moisture, and the vitamin C in them degrades rapidly. They should be stored for as short a time as possible in a cool place, in a container or plastic bag.

## List of Culinary Vegetables

This is a list of vegetables in the culinary sense. This means that the list includes some botanical fruits such as pumpkins, and does not include herbs, spices, cereals and most culinary fruits and culinary nuts. Currently edible fungi are not included on this list.

Some culinary vegetables (seaweeds like laver) are not even members of the plant kingdom.

## Leafy and Salad Vegetables

Iceberg lettuce field in Northern Santa Barbara County.

## Spinach in Flower

- Amaranth (*Amaranthus cruentus*)
- Beet greens (*Beta vulgaris* subsp. *vulgaris*)
- Broccoli Rabe (*Brassica rapa* subsp. *rapa*)
- Bitterleaf (*Vernonia calvoana*)
- Bok choy (*Brassica rapa* Chinensis group)
- Brussels sprout (*Brassica oleracea* Gemmifera group)
- Cabbage (*Brassica oleracea* Capitata group)

- Catsear (*Hypochaeris radicata*)
- Celtuce (*Lactuca sativa* var. *asparagina*)
- Ceylon spinach (*Basella alba*)
- Chicory (*Cichorium intybus*)
- Chinese Mallow (*Malva verticillata*)
- Chrysanthemum leaves (*Chrysanthemum coronarium*)
- Corn salad (*Valerianella locusta*)
- Cress (*Lepidium sativum*)
- Dandelion (*Taraxacum officinale*)
- Endive (*Cichorium endivia*)
- Epazote (*Chenopodium ambrosioides*)
- Fat hen (*Chenopodium album*)
- Fiddlehead (*Pteridium aquilinum, Athyrium esculentum*)
- Fluted pumpkin (*Telfairia occidentalis*)
- Golden samphire (*Inula crithmoides*)
- Good King Henry (*Chenopodium bonus-henricus*)
- Ice plant (*Mesembryanthemum crystallinum*) Jambu (Acmella oleracea)
- Kai-lan (*Brassica rapa* Alboglabra group)
- Kale (*Brassica oleracea* Acephala Group)
- Komatsuna (*Brassica rapa* Pervidis or Komatsuna group)
- Kuka (*Adansonia* sp.)
- Lagos bologi (*Talinum fruticosum*)
- Land cress (*Barbarea verna*)
- Lettuce (*Lactuca sativa*)
- Lizard's tail (*Houttuynia cordata*)
- Melokhia (*Corchorus olitorius, Corchorus capsularis*)
- Mizuna greens (*Brassica rapa* Nipposinica group)

- Mustard (*Sinapis alba*)
- Napa/Chinese Cabbage (*Brassica rapa* Pekinensis group)
- New Zealand Spinach (*Tetragonia tetragonioides*)
- Orache (*Atriplex hortensis*)
- Pea sprouts/leaves (*Pisum sativum*)
- Polk (*Phytolacca americana*)
- Radicchio (*Cichorium intybus*)
- Garden Rocket (*Eruca sativa*)
- Samphire (*Crithmum maritimum*)
- Sea beet (*Beta vulgaris* subsp. *maritima*)
- Seakale (*Crambe maritima*)
- Sierra Leone bologi (*Crassocephalum* spp.)
- Soko (*Celosia argentea*)
- Sorrel (*Rumex acetosa*)
- Spinach (*Spinacia oleracea*)
- Summer purslane (*Portulaca oleracea*)
- Swiss chard (*Beta vulgaris* subsp. *cicla* var. *flavescens*)
- Tatsoi (*Brassica rapa* Rosularis group)
- Turnip greens (*Brassica rapa* Rapifera group)
- Watercress (*Nasturtium officinale*)
- Water spinach (*Ipomoea aquatica*)
- Winter purslane (*Claytonia perfoliata*)
- Yau choy (Brassica napus)

## FRUITING AND FLOWERING VEGETABLES

The fruit of trees:

- Avocado (*Persea americana*)
- Breadfruit (*Artocarpus altilis*)

Fruits of annual or perennial plants:

- Acorn squash (*Cucurbita pepo*)
- Ananas (*Bromeliaceae*)
- Armenian cucumber (*Cucumis melo* Flexuosus group)
- Eggplant or Aubergine (*Solanum melongena*)
- Bell pepper (*Capsicum annuum*)
- Bitter melon (*Momordica charantia*)
- Caigua (*Cyclanthera pedata*)
- Cape Gooseberry (*Physalis peruviana*)
- Cayenne pepper (*Capsicum frutescens*)
- Chayote (*Sechium edule*)
- Chili pepper (*Capsicum annuum* Longum group)
- Cucumber (*Cucumis sativus*)
- Luffa (*Luffa acutangula, Luffa aegyptiaca*)
- Malabar gourd (*Cucurbita ficifolia*)
- Parwal (*Trichosanthes dioica*)
- Perennial cucumber (*Coccinia grandis*)
- Pumpkin (*Cucurbita maxima, Cucurbita pepo*)
- Pattypan squash
- Snake gourd (*Trichosanthes cucumerina*)
- Squash (aka marrow) (*Cucurbita pepo*)
- Sweetcorn aka corn or maize (*Zea mays*)
- Sweet pepper (*Capsicum annuum* Grossum group)
- Tinda (*Praecitrullus fistulosus*)
- Tomato (*Solanum lycopersicum*)
- Tomatillo (*Physalis philadelphica*)
- Winter melon (*Benincasa hispida*)
- West Indian gherkin (*Cucumis anguria*)
- Zucchini or Courgette (*Cucurbita pepo*)

Flowers or flower buds of perennial or annual plants:

- Globe Artichoke (*Cynara scolymus*)
- Squash blossoms (*Cucurbita* spp.)
- Broccoli (*Brassica oleracea*)
- Cauliflower (*Brassica oleracea*)

## Podded Vegetables

Varieties of soybeans are used for many purposes.

- American groundnut (*Apios americana*)
- Azuki bean (*Vigna angularis*)
- Black-eyed pea (*Vigna unguiculata* subsp. *unguiculata*)
- Chickpea (*Cicer arietinum*)
- Drumstick (*Moringa oleifera*)
- Dolichos bean (*Lablab purpureus*)
- Fava bean (*Vicia faba*)
- French bean (*Phaseolus vulgaris*)
- Guar (*Cyamopsis tetragonoloba*)
- Horse gram (*Macrotyloma uniflorum*)
- Indian pea (*Lathyrus sativus*)
- Lentil (*Lens culinaris*)
- Moth bean (*Vigna acontifolia*)
- Mung bean (*Vigna radiata*)
- Okra (*Abelmoschus esculentus*)
- Pea (*Pisum sativum*)
- Peanut (*Arachis hypogaea*)
- Pigeon pea (*Cajanus cajan*)
- Rice bean (*Vigna umbellatta*)
- Runner bean (*Phaseolus coccineus*)
- Soybean (*Glycine max*)

- Tarwi (tarhui, chocho; *Lupinus mutabilis*)
- Tepary bean (*Phaseolus acutifolius*)
- Urad bean (*Vigna mungo*)
- Velvet bean (*Mucuna pruriens*)
- Winged bean (*Psophocarpus tetragonolobus*)
- Yardlong bean (*Vigna unguiculata* subsp. *sesquipedalis*)

## Bulb and Stem Vegetables

Garlic bulbs and individual cloves, one peeled.

- Asparagus (*Asparagus officinalis*)
- Cardoon (*Cynara cardunculus*)
- Celeriac (*Apium graveolens* var. *rapaceum*)
- Celery (*Apium graveolens*)
- Elephant Garlic (*Allium ampeloprasum* var. *ampeloprasum*)
- Florence fennel (*Foeniculum vulgare* var. *dulce*)
- Garlic (*Allium sativum*)
- Kohlrabi (*Brassica oleracea* Gongylodes group)
- Kurrat (*Allium ampeloprasum* var. *kurrat*)
- Leek (*Allium porrum*)
- Lotus root (*Nelumbo nucifera*)
- Nopal (*Opuntia ficus-indica*)
- Onion (*Allium cepa*)
- Prussian asparagus (*Ornithogalum pyrenaicum*)
- Shallot (*Allium cepa* Aggregatum group)
- Welsh onion (*Allium fistulosum*)
- Wild leek (*Allium tricoccum*)

## Root and Tuberous Vegetables

- Ahipa (*Pachyrhizus ahipa*)

- Arracacha (*Arracacia xanthorrhiza*)
- Bamboo shoot
- Beetroot (*Beta vulgaris* subsp. *vulgaris*)
- Black cumin (*Bunium persicum*)
- Burdock (Arctium)
- Broadleaf arrowhead (*Sagittaria latifolia*)
- Camas (*Camassia*)
- Canna (*Canna* spp.)
- Carrot (*Daucus carota*)
- Cassava (*Manihot esculenta*)
- Chinese artichoke (*Stachys affinis*)
- Daikon (*Raphanus sativus* Longipinnatus group)
- Earthnut pea (*Lathyrus tuberosus*)
- Elephant Foot yam (*Amorphophallus paeoniifolius*)
- Ensete (*Ensete ventricosum*)
- Ginger (*Zingiber officinale*)
- Gobo (*Arctium lappa*)
- Hamburg parsley (*Petroselinum crispum* var. *tuberosum*)
- Jerusalem artichoke (*Helianthus tuberosus*)
- Jícama (*Pachyrhizus erosus*)
- Parsnip (*Pastinaca sativa*)
- Pignut (*Conopodium majus*)
- Plectranthus (*Plectranthus* spp.)
- Potato (*Solanum tuberosum*)
- Prairie turnip (*Psoralea esculenta*)
- Radish (*Raphanus sativus*)
- Rutabaga (*Brassica napus* Napobrassica group)

- Salsify (*Tragopogon porrifolius*)
- Scorzonera (*Scorzonera hispanica*)
- Skirret (*Sium sisarum*)
- Sweet Potato (Kumara)
- Taro (*Colocasia esculenta*)
- Ti (*Cordyline fruticosa*)
- Tigernut (*Cyperus esculentus*)
- Turnip (*Brassica rapa* Rapifera group)
- Ulluco (*Ullucus tuberosus*)
- Wasabi (*Wasabia japonica*)
- Water chestnut (*Eleocharis dulcis*)
- Yacón (*Smallanthus sonchifolius*)
- Yam (*Dioscorea* spp.)

## Sea Vegetables

The Caulerpa is a genus of edible Seaweed.

- *Aonori* (*Monostroma* spp., *Enteromorpha* spp.)
- Carola (*Callophyllis variegata*)
- Dabberlocks or badderlocks (*Alaria esculenta*)
- Dulse (*Palmaria palmata*)
- *Hijiki* (*Hizikia fusiformis*)
- *Kombu* (*Laminaria japonica*)
- *Mozuku* (*Cladosiphon okamuranus*)
- *Laver* (*Porphyra* spp.) (nori in Japan, gim in Korea)
- *Ogonori* (*Gracilaria* spp.)
- Sea grape (*Caulerpa* spp.)
- Sea lettuce (*Ulva lactuca*)
- *Wakame* (*Undaria pinnatifida*

# 7

# Fruit

## INTRODUCTION

The term fruit has different meanings dependent on context, and the term is not synonymous in food preparation and biology. In botany, which is the scientific study of plants, fruits are the ripened ovaries of flowering plants. In many plant species, the fruit includes the ripened ovary and surrounding tissues. Fruits are the means by which flowering plants disseminate seeds, and the presence of seeds indicates that a structure is most likely a fruit, though not all seeds come from fruits No single terminology really fits the enormous variety that is found among plant fruits. The term 'false fruit' (pseudocarp, accessory fruit) is sometimes applied to a fruit like the fig (a *multiple-accessory fruit*; see below) or to a plant structure that resembles a fruit but is not derived from a flower or flowers. Some gymnosperms, such as yew, have fleshy arils that resemble fruits and some junipers have *berry-like,* fleshy cones. The term "fruit" has also been inaccurately applied to the seed-containing female cones of many conifers.

### Botanic Fruit and Culinary Fruit

Some vegetables fall into both categories. Like tomatoes etc.

Many true fruits, in a botanical sense, are treated as vegetables in cooking and food preparation because they are not sweet. These botanical fruits include cucurbits (e.g., squash, pumpkin, and cucumber), tomato, peas, beans, corn, eggplant, and sweet pepper, spices, such as allspice and chillies Occasionally, though rarely, a culinary "fruit" is branded as a true fruit in the botanical sense. For example, rhubarb is often referred to as a fruit, because it is used to make sweet desserts such as pies, though only the petiole of the rhubarb plant is edible. In the culinary sense, a fruit is usually any sweet tasting plant product associated with seed(s), a vegetable is any savoury or less sweet plant product, and a nut any hard, oily, and shelled plant product.

Although a nut is a type of fruit, it is also a popular term for edible seeds, such as peanuts (which are actually a legume) and pistachios Technically, a cereal grain is a fruit termed a caryopsis. However, the fruit wall is very thin and fused to the seed coat so almost all of the edible grain is actually a seed. Therefore, cereal grains, such as corn, wheat and rice are better considered edible seeds, although some references list them as fruits Edible gymnosperm seeds are often misleadingly given fruit names, e.g. pine nuts, ginkgo nuts, and juniper berries. A Folk taxonomy is a vernacular naming system which describes how non-scientists categorize items.

## FRUIT DEVELOPMENT

The development sequence of a typical drupe, the nectarine (*Prunus persica*) over a 7½-month period, from bud formation in early winter to fruit ripening in midsummer (see image page for further information)

A fruit is a ripened ovary. Inside the ovary is one or more ovules where the megagametophyte contains the megagamete or egg cell. The ovules are fertilized in a process that starts with pollination, which involves the movement of pollen from the stamens to the stigma of flowers. After pollination, a tube grows from the pollen through the stigma into the ovary to the ovule and sperm are transferred from the pollen to the ovule, within the ovule the sperm unites with the egg, forming a diploid zygote. Fertilization in flowering plants involving both plasmogamy, the fusing of the sperm and egg protoplasm and karyogamy, the union of the sperm and egg nucleus. When the sperm enters the nucleus of the ovule and joins with the megagamete and the endosperm mother cell, the fertilization process is completed As the developing seeds mature, the ovary begins to ripen. The ovules develop into seeds and the ovary wall, the *pericarp*, may become fleshy (as in berries or drupes), or form a hard outer covering (as in nuts).

In some cases, the sepals, petals and/or stamens and style of the flower fall off. Fruit development continues until the seeds have matured. In some multiseeded fruits, the extent to which the flesh develops is proportional to the number of fertilized ovules. The wall of the fruit, developed from the ovary wall of the flower, is called the *pericarp*. The *pericarp* is often differentiated into two or three distinct layers called the *exocarp* (outer layer - also called epicarp), *mesocarp* (middle layer), and *endocarp* (inner layer). In some fruits, especially simple fruits derived from an inferior ovary, other parts of the flower (such as the floral tube, including the petals, sepals, and stamens), fuse with the ovary and ripen with it.

The plant hormone ethylene causes ripening. When such other floral parts are a significant part of the fruit, it is called an *accessory fruit*. Since other parts of the flower may contribute to the structure of the fruit, it is important to study flower structure to understand how a particular fruit forms. Fruits are so diverse that it is difficult to devise a classification scheme that includes all known fruits. Many common terms for seeds and fruit are incorrectly applied, a fact that complicates understanding of the terminology. Seeds are ripened ovules; fruits are the ripened ovaries or carpels that contain the seeds.

To these two basic definitions can be added the clarification that in botanical terminology, a nut is not a type of fruit and not another term for seed, on the contrary to common terminology. There are three basic types of fruits:

— Simple fruit

— Aggregate fruit

— Multiple fruit

Epigynous berries are simple fleshy fruit. From top right: cranberries, lingonberries, blueberries red huckleberries. Simple fruits can be either dry or fleshy, and result from the ripening of a simple or compound ovary with only one pistil. Dry fruits may be either dehiscent (opening to discharge seeds), or indehiscent (not opening to discharge seeds).

Types of dry, simple fruits, with examples of each, are:

- achene - (buttercup, strawberry seeds)
- capsule - (Brazil nut)
- caryopsis - (wheat)
- fibrous drupe - (coconut, walnut)
- follicle - (milkweed)
- legume - (pea, bean, peanut)
- loment
- nut - (hazelnut, beech, oak acorn)
- samara - (elm, ash, maple key)
- schizocarp - (carrot)
- silique - (radish)
- silicle - (shepherd's purse)
- utricle - (beet)

Fruits in which part or all of the *pericarp* (fruit wall) is fleshy at maturity are *simple fleshy fruits*. Types of fleshy, simple fruits (with examples) are:

- berry - (redcurrant, gooseberry, tomato, avocado)
- stone fruit or drupe (plum, cherry, peach, apricot, olive)
- false berry - Epigynous accessory fruits (banana, cranberry, strawberry (edible part)
- pome - accessory fruits (apple, pear, rosehip)

## AGGREGATE FRUIT

Dewberry flowers. Note the multiple pistils, each of which will produce a drupelet. Each flower will become a blackberry-like aggregate fruit.

An aggregate fruit, or *etaerio*, develops from a flower with numerous simple pistils. An example is the raspberry, whose simple fruits are termed *drupelets* because each is like a small drupe attached to the receptacle. In some bramble fruits (such as blackberry) the receptacle is elongated and part of the ripe fruit, making the blackberry an *aggregate-accessory* fruit The strawberry is also an aggregate-accessory fruit, only one in which the seeds are contained in achenes. In all these examples, the fruit develops from a single flower with numerous pistils. Some kinds of aggregate fruits are called berries, yet in the botanical sense they are not.

## MULTIPLE FRUIT

A multiple fruit is one formed from a cluster of flowers (called an *inflorescence*). Each flower produces a fruit, but these mature into a single mass. Examples are the pineapple, edible fig, mulberry, osage-orange, and breadfruit. In some plants, such as this noni, flowers are produced regularly along the stem and it is possible to see together examples of flowering, fruit development, and fruit ripening. In the photograph on the right, stages of flowering and fruit development in the noni or Indian mulberry (*Morinda citrifolia*) can be observed on a single branch. First an inflorescence of white flowers called a head is produced. After fertilization, each flower develops into a drupe, and as the drupes expand, they become *connate* (merge) into a *multiple fleshy fruit* called a *syncarpet*.

There are also many dry multiple fruits, e.g.

- Tuliptree, multiple of samaras.
- Sweet gum, multiple of capsules.
- Sycamore and teasel, multiple of achenes.
- Magnolia, multiple of follicles.

### Common Types of Fruits

The common types of fruit are given below:

- *Berry* — simple fruit and seeds created from a single ovary
- *Pepo* — Berries where the skin is hardened, like cucurbits

- *Hesperidium* — Berries with a rind, like most citrus fruit
- *Epigynous berries (false berries)* — Epigynous fruit made from a part of the plant other than a single ovary.

Compound fruit, which includes:

- *Aggregate fruit* — multiple fruits with seeds from different ovaries of a single flower.
- *Multiple fruit* — fruits of separate flowers, packed closely together.
- *Other accessory fruit* — where the edible part is not generated by the ovary

## Seedless Fruits

An arrangement of fruits commonly thought of as vegetables, including tomatoes and various squash. Seedlessness is an important feature of some fruits of commerce. Commercial cultivars of bananas and pineapples are examples of seedless fruits. Some cultivars of citrus fruits (especially navel oranges and mandarin oranges), table grapes, grapefruit, and watermelons are valued for their seedlessness. In some species, seedlessness is the result of *parthenocarpy*, where fruits set without fertilization. Parthenocarpic fruit set may or may not require pollination. Most seedless citrus fruits require a pollination stimulus; bananas and pineapples do not. Seedlessness in table grapes results from the abortion of the embryonic plant that is produced by fertilization, a phenomenon known as *stenospermocarpy* which requires normal pollination and fertilization.

## Seed Dissemination

Variations in fruit structures largely depend on the mode of dispersal of the seeds they contain. This dispersal can be achieved by animals, wind, water, or explosive dehiscence.

Some fruits have coats covered with spikes or hooked burrs, either to prevent themselves from being eaten by animals or to stick to the hairs, feathers or legs of animals, using them as dispersal agents. Examples include cocklebur and unicorn plant. The sweet flesh of many fruits is "deliberately" appealing to animals, so that the seeds held within are eaten and "unwittingly" carried away and deposited at a distance from the parent. Likewise, the nutritious, oily kernels of nuts are appealing to rodents (such as squirrels) who hoard them in the soil in order to avoid starving during the winter, thus giving those seeds that remain uneaten the chance to germinate and grow into a new plant away from their parent. Other fruits are elongated and flattened out naturally and so become thin, like wings or helicopter blades, e.g. maple, tuliptree and elm. This is an evolutionary mechanism to increase dispersal distance away from the parent via wind. Other wind-dispersed fruit have tiny *parachutes,* e.g. dandelion and salsify. Coconut fruits can float thousands of miles in the ocean to spread seeds. Some other fruits that can disperse via water are nipa palm and screw pine. Some fruits fling seeds substantial distances (up to 100 m in sandbox tree) via explosive dehiscence or other mechanisms, e.g. impatiens and squirting cucumber.

## Uses

Nectarines are one of many fruits that can be easily stewed. Many hundreds of fruits, including fleshy fruits like apple, peach, pear, kiwifruit, watermelon and mango are commercially valuable as human food, eaten both fresh and as jams, marmalade and other preserves. Fruits are also in manufactured foods like cookies, muffins, yoghurt, ice cream, cakes, and many more. Many fruits are used to make beverages, such as fruit juices (orange juice, apple juice, grape juice, etc) or alcoholic beverages, such as wine or brandy. Apples are often used to make vinegar.Fruits are also used for gift giving, Fruit Basket and Fruit Bouquet are some common forms of fruit gifts. Many vegetables are botanical fruits, including tomato, bell pepper, eggplant, okra, squash, pumpkin, green bean, cucumber and zucchini. Olive fruit is pressed for olive oil. Spices like vanilla, paprika, allspice and black pepper are derived from berries.

## Nutritional Value

Fruits are generally high in fiber, water and vitamin C. Fruits also contain various phytochemicals that do not yet have an

RDA/RDI listing under most nutritional factsheets, and which research indicates are required for proper long-term cellular health and disease prevention. Regular consumption of fruit is associated with reduced risks of cancer, cardiovascular disease, stroke, Alzheimer disease, cataracts, and some of the functional declines associated with aging.

**Non-food Uses**

Because fruits have been such a major part of the human diet, different cultures have developed many different uses for various fruits that they do not depend on as being edible. Many dry fruits are used as decorations or in dried flower arrangements, such as unicorn plant, lotus, wheat, annual honesty and milkweed. Ornamental trees and shrubs are often cultivated for their colorful fruits, including holly, pyracantha, viburnum, skimmia, beautyberry and cotoneaster. Fruits of opium poppy are the source of opium which contains the drugs morphine and codeine, as well as the biologically inactive chemical theabaine from which the drug oxycodone is synthysized Osage orange fruits are used to repel cockroaches Bayberry fruits provide a wax often used to make candles Many fruits provide natural dyes, e.g. walnut, sumac, cherry and mulberry.

Dried gourds are used as decorations, water jugs, bird houses, musical instruments, cups and dishes. Pumpkins are carved into Jack-o'-lanterns for Halloween. The spiny fruit of burdock or cocklebur were the inspiration for the invention of Velcro. Coir is a fibre from the fruit of coconut that is used for doormats, brushes, mattresses, floortiles, sacking, insulation and as a growing medium for container plants. The shell of the coconut fruit is used to make souvenir heads, cups, bowls, musical instruments and bird houses.

# 8

# Medicinal Plants

Medicinal plants are plants which may have medicinal properties. Almost all our present medicines come from medicinal plants and are derived from research on medicinal plants.

**"Medicinal Plants"**

- Bamboo
- Bamboo species

**List of Plants Used as Medicine**

**A**

- *Abelmoschus moschatus*
- *Abies balsamea*
- *Abrus precatorius*
- *Acacia berlandieri*
- *Acacia catechu*
- *Acacia chundra*
- *Acacia cornigera*
- *Acacia decurrens*
- *Acacia greggii*
- *Acacia mellifera*

- *Acacia nilotica*
- *Acacia plumosa*
- *Acacia senegal*
- *Acacia sieberiana*
- *Acacia tortilis*
- *Acerola*
- *Achillea millefolium*
- *Acmella oleracea*
- *Acokanthera schimperi*
- *Aconitum carmichaelii*
- *Actaea racemosa*
- *Actinidia chinensis*
- *Adonis aestivalis*
- *Adonis annua*
- *Aegiphila lhotskiana*
- *Aesculus*
- *Aesculus hippocastanum*
- *Aframomum melegueta*
- *Agathosma betulina*
- *Agave americana*
- *Agrimonia eupatoria*
- *Ajuga reptans*
- *Ajwain*
- *Alangium chinense*
- *Albizia julibrissin*
- *Albizia lebbeck*
- *Alchornea*
- *Alchornea castaneifolia*

- *Alchornea cordifolia*
- *Alchornea floribunda*
- *Alepidea peduncularis*
- *Aleppo pepper*
- *Alfalfa*
- *Alhandal*
- *Alisma plantago-aquatica*
- *Allium giganteum*
- *Allium moly*
- *Allspice*
- *Alnus rubra*
- *Aloe*
- *Aloe succotrina*
- *Aloe vera*
- *Alpinia*
- *Alpinia galanga*
- *Alstonia constricta*
- *Althaea (genus)*
- *Althaea officinalis*
- *Amelanchier canadensis*
- *Ammi majus*
- *Anadenanthera colubrina var. cebil*
- *Anadenanthera colubrina var. colubrina*
- *Anadenanthera peregrina var. falcata*
- *Anagallis arvensis*
- *Anamirta cocculus*
- *Andrographis paniculata*
- *Androstephium caeruleum*

- *Anemone*
- *Anemone chinensis*
- *Anemone hepatica*
- *Anemone nemorosa*
- *Anemone ranunculoides*
- *Angelica sinensis*
- *Angostura (genus)*
- *Angostura trifoliata*
- *Anise*
- *Anisodus*
- *Anisodus tanguticus*
- *Anthemis tinctoria*
- *Anthoxanthum odoratum*
- *Anthyllis vulneraria*
- *Aphanes*
- *Apocynum androsaemifolium*
- *Apocynum cannabinum*
- *Apple mint*
- *Aquaretics*
- *Aquilegia grata*
- *Aquilegia vulgaris*
- *Arctostaphylos uva-ursi*
- *Areca catechu*
- *Argemone mexicana*
- *Argentina anserina*
- *Aristolochia*
- *Aristolochia clematitis*
- *Aristolochia grandiflora*

- *Arnica*
- *Arnica montana*
- *Artemisia absinthium*
- *Artemisia afra*
- *Artemisia annua*
- *Artemisia maritima*
- *Artemisia princeps*
- *Artemisia vulgaris*
- *Artichoke*
- *Arum italicum*
- *Arum maculatum*
- *Asafoetida*
- *Asarum*
- *Asarum europaeum*
- *Asarum splendens*
- *Asclepias incarnata*
- *Asclepias verticillata*
- *Ash tree*
- *Asimina triloba*
- *Asparagus*
- *Asparagus racemosus*
- *Aspidosperma quebracho-blanco*
- *Astragalus*
- *Astragalus canadensis*
- *Astragalus propinquus*
- *Astrocaryum aculeatissimum*
- *Asystasia gangetica*

- *Atropa belladonna*
- *Aztekium*
- *Açaí Palm*

**B**

- *Bacopa monnieri*
- *Bael*
- *Balsam poplar*
- *Baptisia australis*
- *Basil*
- *Bay Laurel*
- *Bearberry*
- *Beautyberry*
- *Beet*
- *Belamcanda*
- *Bellis perennis*
- *Berberis vulgaris*
- *Betel*
- *Betula lenta*
- *Bilberry*
- *Bird Cherry*
- *Bitter melon*
- *Bitter orange*
- *Bixa orellana*
- *Black Bryony*
- *Black cardamom*
- *Black locust*
- *Black pepper*
- *Bloodroot*

- *Blueberry*
- *Boesenbergia rotunda*
- *Bois bande*
- *Boldo*
- *Borage*
- *Boswellia*
- *Boxthorn*
- *Boysenberry*
- *Brassica nigra*
- *Bryony*
- *Bunium persicum*
- *Burdock*
- *Burnet Saxifrage*
- *Burututu*
- *Bushweed*
- *Butterfly weed*
- *Buxus sempervirens*

**C**

- *Caesalpinia pulcherrima*
- *Calabar bean*
- *Calea zacatechichi*
- *Calendula officinalis*
- *Callicarpa americana*
- *Callicarpa bodinieri*
- *Callicarpa japonica*
- *Callicarpa macrophylla*
- *Callicarpa nudiflora*

- *Callicarpa pentandra*
- *Caltha palustris*
- *Camellia sinensis*
- *Candlenut*
- *Cannabis*
- *Cannabis sativa*
- *Caper*
- *Capsella bursa-pastoris*
- *Capsicum annuum*
- *Caraway*
- *Cardamine pratensis*
- *Cardamom*
- *Cashew*
- *Castanea pumila*
- *Castor oil plant*
- *Catharanthus*
- *Catharanthus roseus*
- *Caulophyllum*
- *Caulophyllum thalictroides*
- *Cayenne pepper*
- *Cecropia*
- *Cedrus deodara*
- *Celery*

# 9

# Wood

## INTRODUCTION

Wood is an organic material; in the strict sense wood is produced as secondary xylem in the stems of woody plants, notably trees but also shrubs, etc. In a living tree it conducts water and nutrients to the leaves and other growing tissues, and has a support function, enabling plants to reach large sizes. Wood may also refer to other plant materials and tissues with comparable properties, and to material engineered from wood, or wood chips or fibers. People have used wood for millennia for many purposes, primarily as a fuel or as a construction material for making houses, tools, weapons, furniture, packaging, artworks, and paper. Wood can be dated by carbon dating and in some species by dendrochronology to make inferences about when a wooden object was created. The year-to-year variation in tree-ring widths and isotopic abundances gives clues to the prevailing climate at that time

### Formation

A tree increases in diameter by the formation, between the old wood and the inner bark, of new woody layers which envelop the entire stem, living branches, and roots. Where there are clear seasons, this can happen in a discrete pattern, leading to what is known as growth rings, as can be seen on the end of a log. If these seasons are annual these growth rings are annual rings. Where there is no seasonal difference growth rings are likely to be indistinct or absent.

Within a growth ring it may be possible to see two parts. The part nearest the center of the tree is more open textured and almost invariably lighter in colour than that near the outer portion of the ring. The inner portion is formed early in the season, when growth is comparatively rapid; it is known as early wood or spring wood. The outer portion is the late wood or summer wood, being produced in the summer. white pines there is not much contrast in the different parts of the ring, and as a result the wood is very uniform in texture and is easy to work. In hard pines, on the other hand, the late wood is very dense and is deep-colored, presenting a very decided contrast to the soft, straw-colored early wood. In ring-porous woods each season's growth is always well defined, because the large pores of the spring abut on the denser tissue of the fall before. In the diffuse-porous woods, the demarcation between rings is not always so clear and in some cases is almost (if not entirely) invisible to the unaided eye.

## KNOTS

A knot on a tree at the Garden of the Gods public park in Colorado Springs, Colorado. A knot is a particular type of imperfection in a piece of timber, which reduces its strength, but which may be exploited for artistic effect. In a longitudinally-sawn plank, a knot will appear as a roughly circular "solid" (usually darker) piece of wood around which the roughly parallel fibres (grain) of the rest of the "flows" (parts and rejoins). A knot is actually a portion of a side branch (or a dormant bud) included in the wood of the stem or larger branch. The included portion is irregularly conical in shape (hence the roughly circular cross-section) with the tip at the point in stem diameter at which the plant's cambium was located when the branch formed as a bud.

Within a knot, the fibre direction (grain) is up to 90 degrees different from the fibres of the stem, thus producing local cross grain. During the development of a tree, the lower limbs often die, but may persist for a time, sometimes years. Subsequent layers of growth of the attaching stem are no longer intimately joined with the dead limb, but are grown around it. Hence, dead branches produce knots which are not attached, and likely to drop out after the tree has been sawn into boards. In grading lumber and structural timber, knots are classified according to their form, size, soundness, and the firmness with which they are held in place. This firmness is affected by, among other factors, the length of time for which the branch was dead while the attaching stem continued to grow. Knots materially affect cracking (known in the industry as checking) and warping, ease in working, and cleavability of timber. They are defects which weaken timber and lower its value for structural purposes where strength is an important consideration.

The weakening effect is much more serious when timber is subjected to forces perpendicular to the grain and/or tension than where under load along the grain and/or compression. The extent to which knots affect the strength of a beam depends upon their position, size, number, direction of fiber, and condition. A knot on the upper side is compressed, while one on the lower side is subjected to tension. If there is a season check in the knot, as is often the case, it will offer little resistance to this tensile stress. Small knots, however, may be located along the neutral plane of a beam and increase the strength by preventing longitudinal shearing. Knots in a board or plank are least injurious when they extend through it at right angles to its broadest surface.

Knots which occur near the ends of a beam do not weaken it. Sound knots which occur in the central portion one-fourth the height of the beam from either edge are not serious defects. Knots do not necessarily influence the stiffness of structural timber. Only defects of the most serious character affect the elastic limit of beams. Stiffness and elastic strength are more dependent upon the quality of the wood fiber than upon defects in the beam. The effect of knots is to reduce the difference between the fiber stress at elastic limit and the modulus of rupture of beams. The breaking strength is very susceptible to defects. Sound knots do not weaken wood when subject to compression parallel to the grain. For purposes for which appearance is more important than strength, such as wall panelling, knots are considered a benefit, as they add visual texture to the wood, giving it a more interesting appearance. The traditional style of playing the Basque xylophon *txalaparta* involves hitting the right knots to obtain different tones.

## HEARTWOOD AND SAPWOOD

A section of a Yew branch showing 27 annual growth rings, pale sapwood and dark heartwood, and pith (centre dark spot). The dark radial lines are small knots.

Heartwood is wood that has died and become resistant to decay as a result of genetically programmed processes. It appears in a cross-section as a discolored circle, following annual rings in shape. Heartwood is usually much darker than living wood, and forms with age. Many woody plants do not form heartwood, but other processes, such as decay, can discolor wood in similar ways, leading to confusion. Some uncertainty still exists as to whether heartwood is truly dead, as it can still chemically react to decay organisms, but only once.

Sapwood is the wood that is not heartwood. In the growing tree it is living wood. All wood in a tree is first formed as sapwood. Its principal functions are to conduct water from the roots to the leaves and to store up and give back according to the season the food prepared in the leaves. The more leaves a tree bears and the more vigorous its growth, the larger the volume of sapwood required. Hence trees making rapid growth in the open have thicker sapwood for their size than trees of the same species growing in dense forests. Sometimes trees grown in the open may become of considerable size, 30 cm or more in diameter, before any heartwood begins to form, for example, in second-growth hickory, or open-grown pines.

The term *heartwood* derives solely from its position and not from any vital importance to the tree. This is evidenced by the fact that a tree can thrive with its heart completely decayed. Some species begin to form heartwood very early in life, so having only a thin layer of live sapwood, while in others the change comes slowly. Thin sapwood is characteristic of such trees as chestnut, black locust, mulberry, osage-orange, and sassafras, while in maple, ash, hickory, hackberry, beech, and pine, thick sapwood is the rule. Others never form heartwood. There is no definite relation between the annual rings of growth and the amount of sapwood. Within the same species the cross-sectional area of the sapwood is very roughly proportional to the size of the crown of the tree. If the rings are narrow, more of them are required than where they are wide. As the tree gets larger, the sapwood must necessarily become thinner or increase materially in volume.

Sapwood is thicker in the upper portion of the trunk of a tree than near the base, because the age and the diameter of the upper sections are less. When a tree is very young it is covered with limbs almost, if not entirely, to the ground, but as it grows older some or all of them will eventually die and are either broken off or fall off. Subsequent growth of wood may completely conceal the stubs which will however remain as knots. No matter how smooth and clear a log is on the outside, it is more or less knotty near the middle. Consequently the sapwood of an old tree, and particularly of a forest-grown tree, will be freer from knots than the inner heartwood.

Since in most uses of wood, knots are defects that weaken the timber and interfere with its ease of working and other properties, it follows that a given piece of sapwood, because of its position in the tree, may well be stronger than a piece of heartwood from the same tree. It is remarkable that the inner heartwood of old trees remains as sound as it usually does, since in many cases it is hundreds of years, and in a few instances thousands of years, old. Every broken limb or root, or deep wound from fire, insects, or falling timber, may afford an entrance for decay, which, once started, may penetrate to all parts of the trunk. The larvae of many insects bore into the trees and their tunnels remain indefinitely as sources of weakness. Whatever advantages, however, that sapwood may have in this connection are due solely to its relative age and position.

If a tree grows all its life in the open and the conditions of soil and site remain unchanged, it will make its most rapid growth in youth, and gradually decline. The annual rings of growth are for many years quite wide, but later they become narrower and narrower. Since each succeeding ring is laid down on the outside of the wood previously formed, it follows that unless a tree materially increases its production of wood from year to year, the rings must necessarily become thinner as the trunk gets wider. As a tree reaches maturity its crown becomes more open and the annual wood production is lessened, thereby reducing still more the width of the growth rings. In the case of forest-grown trees so much depends upon the competition of the trees in their struggle for light and nourishment that periods of rapid and slow growth may alternate.

Some trees, such as southern oaks, maintain the same width of ring for hundreds of years. Upon the whole, however, as a tree gets larger in diameter the width of the growth rings decreases. There may be decided differences in the grain of different pieces of wood cut from a large tree, particularly one that is mature. In some trees, the wood laid on late in the life of a tree is softer, lighter, weaker, and more even-textured than that produced earlier, but in other trees, the reverse applies. This may or may not correspond to heartwood and sapwood. In a large log the sapwood, because of the time in the life of the tree when it was grown, may be inferior in hardness, strength, and toughness to equally sound heartwood from the same log.

## DIFFERENT WOODS

There is a strong relationship between the properties of wood and the properties of the particular tree that yielded it. For every tree species there is a range of density for the wood it yields. There is a rough correlation between density of a wood and its strength (mechanical properties). For example, while mahogany is a medium-dense hardwood which is excellent for fine furniture crafting, balsa is light, making it useful for model building.

The densest wood may be black ironwood. Wood is commonly classified as either softwood or hardwood. The wood from conifers (e.g. pine) is called softwood, and the wood from broad-leaved trees (e.g. oak) is called hardwood. These names are a bit misleading, as hardwoods are not necessarily hard, and softwoods are not necessarily soft. The well-known balsa (a hardwood) is actually softer than any commercial softwood.

Conversely, some softwoods (e.g. yew) are harder than most hardwoods. Wood products such as plywood are typically classified as engineered wood and not considered raw wood.

## Colour

In species which show a distinct difference between heartwood and sapwood the natural colour of heartwood is usually darker than that of the sapwood, and very frequently the contrast is conspicuous. This is produced by deposits in the heartwood of various materials resulting from the process of growth, increased possibly by oxidation and other chemical changes, which usually have little or no appreciable effect on the mechanical properties of the wood. Some experiments on very resinous Longleaf Pine specimens, however, indicate an increase in strength. This is due to the resin which increases the strength when dry. Such resin-saturated heartwood is called "fat lighter". Structures built of fat lighter are almost impervious to rot and termites; however they are very flammable. Stumps of old longleaf pines are often dug, split into small pieces and sold as kindling for fires. Stumps thus dug may actually remain a century or more since being cut.

Spruce impregnated with crude resin and dried is also greatly increased in strength thereby. The wood of Coast Redwood is distinctively red in colour. Since the late wood of a growth ring is usually darker in colour than the early wood, this fact may be used in judging the density, and therefore the hardness and strength of the material. This is particularly the case with coniferous woods. In ring-porous woods the vessels of the early wood not infrequently appear on a finished surface as darker than the denser late wood, though on cross sections of heartwood the reverse is commonly true. Except in the manner just stated the colour of wood is no indication of strength. Abnormal discolouration of wood often denotes a diseased condition, indicating unsoundness. The black check in western hemlock is the result of insect attacks. The reddish-brown streaks so common in hickory and certain other woods are mostly the result of injury by birds. The discolouration is merely an indication of an injury, and in all probability does not of itself affect the properties of the wood. Certain rot-producing fungi impart to wood characteristic colours which thus become symptomatic of weakness; however an attractive effect known as spalting produced by this process is often considered a desirable characteristic. Ordinary sap-staining is due to fungous growth, but does not necessarily produce a weakening effect.

## STRUCTURE

Wood is a heterogeneous, hygroscopic, cellular and anisotropic material. It is composed of fibers of cellulose (40%-50%) and hemicellulose (15%-25%) impregnated with lignin (15%-30%). In coniferous or softwood species the wood cells are mostly of one kind, tracheids, and as a result the material is much more uniform in structure than that of most hardwoods. There are no vessels ("pores") in coniferous wood such as one sees so prominently in oak and ash, for example. The structure of the hardwoods is more complex.

They are more or less filled with vessels: in some cases (oak, chestnut, ash) quite large and distinct, in others (buckeye, poplar, willow) too small to be seen plainly without a small hand lens. In discussing such woods it is customary to divide them into two large classes, *ring-porous* and *diffuse-porous*. In ring-porous species, such as ash, black locust, catalpa, chestnut, elm, hickory, mulberry, and oak, the larger vessels or pores (as cross sections of vessels are called) are localized in the part of the growth ring formed in spring, thus forming a region of more or less open and porous tissue. The rest of the ring, produced in summer, is made up of smaller vessels and a much greater proportion of wood fibres. These fibres are the elements which give strength and toughness to wood, while the vessels are a source of weakness.

In diffuse-porous woods the pores are scattered throughout the growth ring instead of being collected in a band or row. Examples of this kind of wood are basswood, birch, buckeye, maple, poplar, and willow. Some species, such as walnut and cherry, are on the border between the two classes, forming an intermediate group. If a heavy piece of pine is compared with a light specimen it will be seen at once that the heavier one contains a larger proportion of late wood than the other, and is therefore considerably darker..

The late wood of all species is denser than that formed early in the season, hence the greater the proportion of late wood the greater the density and strength. When examined under a microscope the cells of the late wood are seen to be very thick-walled and with very small cavities, while those formed first in the season have thin walls and large cavities. The strength is in the walls, not the cavities. In choosing a piece of pine where strength or stiffness is the important consideration, the principal thing to observe is the comparative amounts of early and late wood. The width of ring is not nearly so important as the proportion of the late wood in the ring. It is not only the proportion of late wood, but also its quality, that counts. In specimens that show a very large proportion of late wood it may be noticeably more porous and weigh considerably less than the late wood in pieces that contain but little.

One can judge comparative density, and therefore to some extent weight and strength, by visual inspection. No satisfactory explanation can as yet be given for the real causes underlying the formation of early and late wood. Several factors may be involved. In conifers, at least, rate of growth alone does not determine the proportion of the two portions of the ring, for in some cases the wood of slow growth is very hard and heavy, while in others the opposite is true. The quality of the site where the tree grows undoubtedly affects the character of the wood formed, though it is not possible to formulate a rule governing it. In general, however, it may be said that where strength or ease of working is essential, woods of moderate to slow growth should be chosen.

But in choosing a particular specimen it is not the width of ring, but the proportion and character of the late wood which should govern. In the case of the ring-porous hardwoods there seems to exist a pretty definite relation between the rate of growth of timber and its properties. This may be briefly summed up in the general statement that the more rapid the growth or the wider the rings of growth, the heavier, harder, stronger, and stiffer the wood. This, it must be remembered, applies only to ring-porous woods such as oak, ash, hickory, and others of the same group, and is, of course, subject to some exceptions and limitations. In ring-porous woods of good growth it is usually the middle portion of the ring in which the thick-walled, strength-giving fibers are most abundant.

As the breadth of ring diminishes, this middle portion is reduced so that very slow growth produces comparatively light, porous wood composed of thin-walled vessels and wood parenchyma. In good oak these large vessels of the early wood occupy from 6 to 10 per cent of the volume of the log, while in inferior material they may make up 25 per cent or more. The late wood of good oak, except for radial grayish patches of small pores, is dark colored and firm, and consists of thick-walled fibers which form one-half or more of the wood. In inferior oak, such fiber areas are much reduced both in quantity and quality. Such variation is very largely the result of rate of growth. Wide-ringed wood is often called "second-growth", because the growth of the young timber in open stands after the old trees have been removed is more rapid than in trees in the forest, and in the manufacture of articles where strength is an important consideration such "second-growth" hardwood material is preferred.

This is particularly the case in the choice of hickory for handles and spokes. Here not only strength, but toughness and resilience are important. The results of a series of tests on hickory by the U.S. Forest Service show that:

> "The work or shock-resisting ability is greatest in wide-ringed wood that has from 5 to 14 rings per inch (rings 1.8-5 mm thick), is fairly constant from 14 to 38 rings per inch (rings 0.7-1.8 mm thick), and decreases rapidly from 38 to 47 rings per inch (rings 0.5-0.7 mm thick). The strength at maximum load is not so great with the most rapid-growing wood; it is maximum with from 14 to 20 rings per inch (rings 1.3-1.8 mm thick), and again becomes less as the wood becomes more closely ringed. The natural deduction is that wood of first-class mechanical value shows from 5 to 20 rings per inch (rings 1.3-5 mm thick) and that slower growth yields poorer stock. Thus the inspector or buyer of hickory

should discriminate against timber that has more than 20 rings per inch (rings less than 1.3 mm thick). Exceptions exist, however, in the case of normal growth upon dry situations, in which the slow-growing material may be strong and tough."

The effect of rate of growth on the qualities of chestnut wood is summarized by the same authority as follows:

"When the rings are wide, the transition from spring wood to summer wood is gradual, while in the narrow rings the spring wood passes into summer wood abruptly. The width of the spring wood changes but little with the width of the annual ring, so that the narrowing or broadening of the annual ring is always at the expense of the summer wood. The narrow vessels of the summer wood make it richer in wood substance than the spring wood composed of wide vessels. Therefore, rapid-growing specimens with wide rings have more wood substance than slow-growing trees with narrow rings. Since the more the wood substance the greater the weight, and the greater the weight the stronger the wood, chestnuts with wide rings must have stronger wood than chestnuts with narrow rings. This agrees with the accepted view that sprouts (which always have wide rings) yield better and stronger wood than seedling chestnuts, which grow more slowly in diameter."

In diffuse-porous woods, as has been stated, the vessels or pores are scattered throughout the ring instead of collected in the early wood. The effect of rate of growth is, therefore, not the same as in the ring-porous woods, approaching more nearly the conditions in the conifers. In general it may be stated that such woods of medium growth afford stronger material than when very rapidly or very slowly grown. In many uses of wood, strength is not the main consideration. If ease of working is prized, wood should be chosen with regard to its uniformity of texture and straightness of grain, which will in most cases occur when there is little contrast between the late wood of one season's growth and the early wood of the next.

## MONOCOT WOOD

Trunks of the Coconut palm, a monocot, in Java. From this perspective these look not much different from trunks of a dicot or conifer. Structural material that roughly (in its gross handling characteristics) resembles ordinary, 'dicot" or conifer wood is produced by a number of monocot plants, and these are also usually called wood. Of these, bamboo, botanically a member of the grass family, has considerable economic importance, larger culms being widely used as a building and construction material in their own right and, these days, in the manufacture of engineered flooring, panels and veneer. Another major plant group that produce material that often is called wood are the palms. Of much less importance are plants such as *Pandanus*, *Dracaena* and *Cordyline*. With all this material, the structure and composition of the structural material is quite different from ordinary wood.

### Water Content

The churches of Kizhi, Russia are among a handful of World Heritage Sites built entirely of wood, without metal joints. Water occurs in living wood in three conditions, namely: (1) in the cell walls; (2) in the protoplasmic contents of the cells; and (3) as free water in the cell cavities and spaces. In heartwood it occurs only in the first and last forms. Wood that is thoroughly air-dried retains from 8-16% of water in the cell walls, and none, or practically none, in the other forms. Even oven-dried wood retains a small percentage of moisture, but for all except chemical purposes, may be considered absolutely dry. The general effect of the water content upon the wood substance is to render it softer and more pliable. A similar effect of common observation is in the softening action of water on paper or cloth. Within certain limits, the greater the water content, the greater its softening effect.

Drying produces a decided increase in the strength of wood, particularly in small specimens. An extreme example is the case of a completely dry spruce block 5 cm in section, which will sustain a permanent load four times as great as that which a green (undried) block of the same size will support. The greatest increase due to drying is in the ultimate crushing strength, and strength at elastic limit in endwise compression; these are followed by the modulus of rupture, and stress at elastic limit in cross-bending, while the modulus of elasticity is least affected.

## Uses

### *Fuel*

Wood is burned as a fuel mostly in rural areas of the world. Hard wood is preferred over softwood because it creates less smoke and burns longer. Adding a woodstove or fireplace to a home adds ambiance and warmth

### Construction

Wood can be cut into straight planks and made into a hardwood floor (parquetry).

The Saitta House, Dyker Heights, Brooklyn, New York built in 1899 is made of and decorated in wood. Wood has been an important construction material since humans began building shelters, houses and boats. Nearly all boats were made out of wood until the late 19th century, and wood remains in common use today in boat construction. New domestic housing in many parts of the world today is commonly made from timber-framed construction. In buildings made of other materials, wood will still be found as a supporting material, especially in roof construction, in interior doors and their frames, and as exterior cladding. Wood to be used for construction work is commonly known as *lumber* in North America. Elsewhere, *lumber* usually refers to felled trees, and the word for sawn planks ready for use is *timber*.

Wood is also commonly used as shuttering material to form the mould into which concrete is poured during reinforced concrete construction. Wood unsuitable for construction in its native form may be broken down mechanically (into fibres or chips) or chemically (into cellulose) and used as a raw material for other building materials such as chipboard, engineered wood, hardboard, medium-density fiberboard (MDF), oriented strand board (OSB). Such wood derivatives are widely used: wood fibers are an important component of most paper, and cellulose is used as a component of some synthetic materials. Wood derivatives can also be used for kinds of flooring, for example laminate flooring. Wood is also used for cutlery, such as chopsticks, toothpicks, and other utensils, like the wooden spoon.

## WOOD ECONOMY

The existence of a wood economy, or more correctly, a forest economy (since in many countries a bamboo economy predominates), is a prominent matter in many developing countries as well as in many other nations with temperate climate and especially in those with low temperatures. These are generally the countries with greater forested areas. The uses of wood in furniture, buildings, bridges, and as a source of energy are widely known. Additionally, wood from trees and bushes, can be employed in a wide variety, including those produced from wood pulp, as cellulose in paper, celluloid in early photographic film, cellophane, and rayon (a substitute for silk).

At the end of their normal useage, wood products can be burnt to obtain thermal energy, or can be used as a fertilizer. The potential environmental damage that a wood economy could occasion include (problems of reduction the biodiversity due to monoculture forestry - the intensive cultivation of very few types of trees); and $CO_2$ emissions. However, forests can aid in reduction of atmospheric carbon dioxide and therefore diminish global warming.

### History of Use of Wood

The wood economy is historically the starting point of the civilizations worldwide, since eras preceding the Paleolithic and the Neolithic. It necessarily preceded [Bronze age] ages of metals by many centuries, as the melting of metals was possible only through the discovery of techniques to light fire (usually obtained by the scraping of two very dry wooden rods) and the building of many simple machines and rudimentary tools, as canes, club handles, bows, arrows, lances. One of the most ancient handmade articles ever found is one smoothed pricked of wood (Clacton Spear) 250,000 years old (third interglacial period), that was buried under sediments in England, at Clacton-on-Sea, and it was probably manifactured by the human-like species Homo erectus. Successive civilizations such as the Egyptians and Sumerians built sophisticated objects of furniture. Many types of furniture in ivory and valuable woods have survived to our time practically intact, because secluded in inviolated secret tombs, they were protected from decay also by the dry environment of desert. Many buildings and parts of these (above all roofs) contained elements in wood (often of oak) forming structural supports and covering; means of transport such as boats, ships; and later (with the invention of the wheel) wagons and carriages, winches, flour mills powered by water, etc.

## Dimensions and Geography of Wood Economy

The main source of the lumber used in the world are forests, which can be classified in virgin, semivirgin and plantation. It is known that much timber is removed for firewood by the local populations in many countries, especially in third world, but this amount can only estimated, with wide margins of uncertainty. In 1998, the world-wide production of wood, officially that not used as firewood (known as "roundwood"), was about 1.5 billions of cubic meters ($m^3$), amounting to around 45% of the wood cultivated in the world. Cut logs and branches destined to become elements for the construction of buildings account for approximately 55% of the World's industrial wood production. 25% becomes wood pulp (including wood powder and truccioli) mainly destined for the production of paper and cardboard; a further 20% approximately became panels in plywood and valuable wood for furnitures and objects of common use (FAO 1998) . The World's largest producer and consumer of this wood "officially accounted" is the USA, although the the country that possesses the greatest area of forests is Russia.

In the seventies, the countries with the biggest forest surface were: Soviet Union (approximately 8,800,000 $km^2$), Brazil (5,150,000 $km^2$), Canada (4,400,000 $km^2$), USA (3,000,000 $km^2$), Indonesia (1,200,000 $km^2$) and Democratic Republic of Congo (1,000,000 $km^2$). Other countries with important production and consumption of wood usually have a low density of population in relation to their territorial extension, here we can include countries as Argentina, Chile, Finland, Poland, Sweden, Ukraine.

By 2001 the rainforest areas of Brazil were reduced by a fifth (respect of 1970), to around 4,000,000 km²; the ground cleared was mainly destined for cattle pasture - Brazil is the world's largest exporter of beef with almost 200,000,000 head of cattle Also the booming ethanol economy that in Brazil is based upon sugar cane cultivation, is reducing forests size. Canadian forest was reduced to by almost 30% to 3,101,340 km² over the same period

## Importance in Fighting Greenhouse Effect

Regarding the problem of climate change, it is known that burning forests increase $CO_2$ in atmosphere, while intact virgin forest or plantations act as sinks for $CO_2$, for these reasons wood economy fights greenhouse effect. The amount of $CO_2$ absorbed depends on the type of trees, lands and the climate of the place where trees naturally grow or are planted. Moreover, should be told that by night plants do not execute photosynthesis, and produce $CO_2$, eliminated the successive day. Paradoxically in summer oxygen created by photosynthesis in forests near to cities and urban parks, interacts with urban air pollution (from cars, etc.) and is transformed by solar beams in ozone (molecule of three oxygen atoms), that while in high atmosphere constitutes a filter against ultraviolet beams, in the low atmosphere is a pollutant, able to provoke respiratory disturbances. In a low-carbon economy, forestry operations will be focused on low-impact practices and regrowth. Forest managers will make sure that they do not disturb soil based carbon reserves too much. Specialized tree farms will be the main source of material for many products. Quick maturing tree varieties will be grown on short rotations in order to maximize output.

## Wood Economy in Australia

*Eucalyptus:* these are seven hundred tree species from Australia, that grow very fast in tropical, sub-tropical and semi-arid climates, and are very resistant to forest fires (with their tree cortex) and to drought. Its essential oil is used in pharmacology, its wood for building, and the small branches as firewood and for paper and cardboard pulp.

## Wood Economy in Brazil

Brazil has a long tradition in the harvesting of several types of trees with specific uses. Since the sixties imported species of pine tree and eucalyptus have been grown mostly for the industry of paper pulp and plywood. Currently high-level research is being conducted, to apply the enzymes of sugar cane fermentation to cellulose in wood, in order to obtain methanol.

*Brazilwood:* has a dense, orange-red heartwood that takes a high red shine (brasa=ember), and it is the premier wood used for making bows for string instruments from the violin family. This trees soon became the biggest source of red dye, and they were such a large part of the economy and export of that country, that slowly it was known as Brazil.

*Hevea brasiliensis:* is the biggest source of the best latex, that is used to manufacture many objects in rubber, as an example gloves, condoms, anti-allergic mattresses and tires (vulcanized rubber). Latex has the ability to adjust to the exact shape of the body part, an advantage over polyurethane or polyethylene gloves.

## Wood Economy in Canada and USA

There is a close relation in the forestry economy between these countries, they have many tree genus in common, and Canada is the main producer of wood and wooden items destined to the USA, the biggest consumer of wood and it's byproducts in the world. The water systems of the Great Lakes, Erie canal, Hudson river and Saint Lawrence Seaway to the east coast and Mississippi to central plains and Louisiana allows transportation of logs at very low costs.

### Canada

The agency Canada Wood Council calculates that by the year 1999 in Canada, the forest sector employed 850,000 workers (1 job every 17), making around $74 billion of value in goods and services. For many years products derived from trees in Canadian forests had been the most important export items of the country. In 2001, exports around the world totaled some $44.1 billion – the single largest contributor to Canadian trade balance. Canada is the world leader in sustainable forestry management practices. Only 120 million ha (28% of Canadian forests) are currently managed for timber production while an estimated 32 million ha are protected from harvesting by the current legislation.

## USA

*Cherry:* has a hardwood prized for its high quality in grain, width, color, and rich warm glow The first trees were carried to the lands surrounding Rome (Latium) from Armenia. In the United States, most cherry trees are grown in Washington, Pennsylvania, West Virginia, California and Oregon. When cherry tree flower, they make several among the most appreciated landscapes in Japan.

*Cedar:* this genus are coniferas of the pinaceae family, originating from high mountain areas from Carpathians, Lebanon, Turkey to Himalaya. Their scented wood make them suitable for chests and closet lining. Cedar oil and wood is known to be a natural repellent to moths Actually are planted in western and southern USA, mostly for ornamental purposes, but also for the production of pencils (specially incense-cedar).

*Douglas-fir:* is a native tree of the United States west coast, with records in fast grow and the reaching of high statures in brief time. It has the ability to grow in mountains till the height of 1,800 meters. Their wood is used for construction, for homebuilt aircraft, for paper pulp, and also as firewood.

*Walnut:* It is a prized furniture and carving hardwood because of its colour, hardness, grain and durability. Walnut wood has been the timber of choice for gun makers for centuries, including the Lee Enfield rifle. It remains one the most popular choices for rifle and shotgun stocks.

## Wood Economy in the Caribbean and Central America

*Mahogany:* has a straight grain, usually free of voids and pockets. The most prized species come from Cuba and Honduras. It has a reddish-brown color, which darkens over time, and displays a beautiful reddish sheen when polished. It has excellent workability, is available in big boards, and is very durable.

Mahogany is used in the making of many musical instruments, as drums, acoustic and electric guitars' back and side, and luxury headphones.

## WOOD ECONOMY IN EUROPE

### Italy

The species that are ideal for the many uses in this type of economy are those employed by arboriculture, that are very well known for their features and the need for certain types of ground and climates.

*Fraxinus:* being a lightweight wood is easy to transport, as firewood burns easily, grows in damp environments like those present in river flooding areas, stands pollution of water and air.

*Larix:* in Italy it grows at high altitudes around mountain tops, it's timber stand sudden climatic change, from icy winds to high temperatures in sunny afternoon summers, it's excellent to be used for the building of exposed structures as bridges, roofs, etc.

*Stone pine:* "Mediterranean pine" could be the noble emblem of many coastal areas in Italy, originally giant forests of pines extended from the mouth of the Tiber river until Liguria and Provence in France, over soils with high salinity, not very apt for agriculture. It's trees produce a vast amount of dry branches that can be burnt, cones (used for Christmas decoration) and needle-like foliage that can be burnt, or used as mulch. Oils and resins can be used in scents and ointments. The pinoli are useful elements in Italian cooking (along with basil are tritured to make pesto sauce). Currently, "progress" has brought to a severe reduction of this magnificent tree extensions, and in many places cheap beach buildings, car-parking and semi-abandoned areas have taken their place.

*Poplar:* in Italy is the most important species for tree plantations, is used for several purposes as plywood manufacture, packing boxes, paper, matches, etc. It needs good quality grounds with good drainage, but can be used to protect the cultivations if disposed in windbreak lines. More than 70% of Italian poplar cultivations are located in the pianura Padana. Constantly the extension of the cultivation is being reduced, from 650 $km^2$ in the 80's to current 350 $km^2$. The yield of poplars is about 1,500 t/$km^2$ of wood every year The production from poplars is around 45-50% of the total italian wood production.

In the history of art poplar was the wood of choice for painting surfaces as panels, as in Renaissance (The Mona Lisa by Leonardo da Vinci). Because of this reason, many of the products with the highest added value, extremely expensive, are made with wood from the humble but durable poplar. Because of the presence of tannic acid, poplar cortex was often used in Europe for the tanning of leather.

## Portugal

*Oak for cork:* are trees with a slow growth, but long life, are cultivated in warm hill areas (min. temp. > -5°Celsius) in all the west area of Mediterranean shores. Popular for bulletin boards. Even if the production as stoppers for wine bottles is diminishing in favor of nylon stoppers, in the sake of energy saving granules of cork can be mixed into concrete. This composites have low thermal conductivity, low density and good energy absorption (earthquake resistant). Some of the property ranges of the composites are density (400–1500 kg/$m^3$), compressive strength (1–26 MPa) and flexural strength (0.5–4.0 MPa . Because of this cork can be used as thermal isolation in buildings (as well in it's natural form and as a mixture), useful also as sound insulation. In the shoe industry cork is used as soles and insoles. In the world there are 20,000 $km^2$ of cork oak plantations, and every year are extracted around 300,000 tons of cork, 50% in Portugal, 15,000 in Italy (12,000 in the island of Sardinia). The advantage of this natural industry is that the extraction of cork from layers outer to the cortex does not kills the tree.

## Wood Economy in Scandinavia and Russia

*Birch:* is a genus with many species of trees from Scandinavia and Russia, excellent for acid grounds. They act as pioneer species in the frozen border between taiga and tundra, are very resistant to periods of drought and icy conditions. The species Betula nana has been identified as the ideal tree for the acid grounds of the sides of sloped mountains, also in southern Europe, with soils poor in nutrients, where these trees can be used to restraint landslides. From birch tree can be extracted Xylitol, a natural sweetener.

## USES OF WOOD

### Combustion

Mean energy density of Wood, was calculated at around 6-17 Megajoule/kilogram.

### Charcoal

#### *Direct Burning*

The most intuitive use for wood, is that for burning in traditional fireplaces, even if combustion of wood is linked to the production of micro-environmental pollutants, as carbon dioxide ($CO_2$), and carbon monoxide (CO), an invisible gas able to provoke irreversible saturation of blood's hemoglobine, as well as nanoparticles.

In Italy poplar has been proposed as a tree cultivation to be transformed in biofuels, interesting because of the excellent ratio between the low energy needed to cultivate, cut and transport these trees, and the high amount of energy that can be extracted from it's wood, because poplar's capture of atmospheric carbon dioxide and their fast growing. As an example, Populus euroamericana clone "I-214", grows so fast that is able to reach 14 inches (35 cm) in diameter and heights of 100 feet (30 m) in ten years.

#### *Wood Gasogen*

Wood gas generator (gasogen): is a bulky and heavy device (but technically simple) that transforms burning wood in a mix of molecular hydrogen ($H_2$), carbon monoxide (CO), carbon dioxide ($CO_2$), molecular nitrogen ($N_2$) and water vapor ($H_2O$). This gas mixture, known as "poor gas" or "syngas" is obtained after the combustion of dry wood in a reductive environment (low in oxygen) with a limited amount of atmospheric air, at temperatures of 900° Celsius, through an internal combustion engine.

A car built in the forties by Ilario Bandini, with a wood gas generator device.In the time between World War I and World War II included, because of the lack of oil, in many countries, like Italy, France, Great Britain and Sweden, several gasoline-powered cars were modified, with the addition of a wood gas generator (a "gasogen"), a device powered by wood, coal, or burnable waste, able to produce (and purify) gas that immediately, in the same vehicle, could power a slightly modified ICE engine of a standard car (low-compression engine). Carburetor had to be changed with an air-gas mixer). There were several setbacks, as the great reduction of maximum speed and the need to drive using low gears and wisely dosing the amount of air. In modern cars, modified with a wood gas generator, gas emissions (CO, $CO_2$ and $NO_x$) are lower to those of the same vehicle running with gasoline (keeping the same catalytic converter).

## Methanol

Methanol, also known as formic alcohol (H3C-OH), is the simplest alcohol (an hydroxyl radical bonded to a molecule of methane), which behaves as a liquid at 25° C, is very toxic (lethal) and corrosive, and in organic chemistry basic books is often called "the spirit of wood", since it can be obtained from wood fermentation. Rarely, when unwise wine-makers mix small chunks of wood and leaves with grapes, methanol can be found as a pollutant of the blend of water, ethanol and other substances derived from grape's fermentation.

Best way to obtain methanol from wood is through syngas (CO, $CO_2$, $H_2$) produced by the anhydrous pyrolysis of wood, a method discovered by ancient egyptians.

Mthanol can be used as an oxygen-rich additive for gasoline, but ussually it is much cheaper to produce it from methane or from syngas, and it is the most important base material for industrial chemistry, where it is often used to make more complex molecules, through reactions of halogenation and later by chemical addition reaction.

## Gas Turbine

The american main battle tank M1 Abrams is powered by a gas turbine of 1,500 hp (1,100 kW), that it is able to function also with a mix at 50% of wood powder and biodiesel, diesel fuel or kerosene. It's advantages over turbo-diesel engine, are the small size and light weight, the lack of a radiator (which gives an advantage against the effect of gun and cannon shots and missile strikes suffered in battle). A setback is the high fuel consumption, since the turbine engine has not the ability to work at a low revolutions per minute rate, much lower than ideal, and during the march this engine consumes twice as much fuel as a modern turbo-diesel engine with intercooler and direct injection.

## CONSTRUCTION

### Structural Wood

Wood is relatively light in weight, because it's specific weight is less than 500 kg/m³, this is an advantage, when compared against 2,000-2,500 kg/m³ for armed concrete or 7,800 kg/m³ for steel. Is strong, becausee the efficiency of wood for structural purposes has qualities that are similar to steel. Wood is used to build bridges (as the Magere bridge in Amsterdam), as well as water and air mills, and microhydro generators for electricity.

### Housing

Structural wood is used as a resource for the building of houses, churches with a broad range of dimensions.

# 10

# Cotton

## INTRODUCTION

Cotton is a soft, staple fiber that grows around the seeds of the cotton plant (*Gossypium* sp.), a shrub native to tropical and subtropical regions around the world, including the Americas, India and Africa. The fiber most often is spun into yarn or thread and used to make a soft, breathable textile, which is the most widely used natural-fiber cloth in clothing today. It is a natural fibre. The English name, which began to be used circa 1400, derives from the Arabic *(al)qutn* meaning cotton. In the 19th and early 20th centuries, In the Southern United States, cotton was known as "King Cotton" because of the great economic and cultural influence it had there. Cotton plants as imagined and drawn by John Mandeville in the fourteenth century. Cotton was cultivated by the inhabitants of the Indus Valley Civilization by the 5th millennium BCE - 4th millennium BCE. The Indus cotton industry was well developed and some methods used in cotton spinning and fabrication continued to be used till the modern Industrialization of India.

Well before the Common Era the use of cotton textiles had spread from India to the Mediterranean and beyond. According to *The Columbia Encyclopedia,* (Sixth Edition):

> "Cotton has been spun, woven, and dyed since prehistoric times. It clothed the people of ancient India, Egypt, and

China. Hundreds of years before the Christian era cotton textiles were woven in India with matchless skill, and their use spread to the Mediterranean countries. In the 1st cent. Arab traders brought fine muslin and calico to Italy and Spain. The Moors introduced the cultivation of cotton into Spain in the 9th cent. Fustians and dimities were woven there and in the 14th cent. in Venice and Milan, at first with a linen warp. Little cotton cloth was imported to England before the 15th cent., although small amounts were obtained chiefly for candlewicks. By the 17th cent. the East India Company was bringing rare fabrics from India. Native Americans skillfully spun and wove cotton into fine garments and dyed tapestries. Cotton fabrics found in Peruvian tombs are said to belong to a pre-Inca culture. In color and texture the ancient Peruvian and Mexican textiles resemble those found in Egyptian tombs."

The earliest cultivation of cotton discovered thus far in the Americas occurred in Mexico, some 5,000 years ago. The indigenous species was *Gossypium hirsutum* which is today the most widely planted species of cotton in the world, constituting about 90% of all production worldwide. The greatest diversity of wild cotton species is found in Mexico, followed by Australia and Africa. In Peru, cultivation of the indigenous cotton species *Gossypium barbadense* was the backbone of the development of coastal cultures such as the Norte Chico, Moche and Nazca.

Cotton was grown upriver, made into nets and traded with fishing villages along the coast for large supplies of fish. The Spanish who came to Mexico in the early 1500s found the people growing cotton and wearing clothing made of it. During the late medieval period, cotton became known as an imported fiber in northern Europe, without any knowledge of how it was derived, other than that it was a plant; noting its similarities to wool, people in the region could only imagine that cotton must be produced by plant-borne sheep. John Mandeville, writing in 1350, stated as fact the now-preposterous belief: "There grew there [India] a wonderful tree which bore tiny lambs on the endes of its branches. These branches were so pliable that they bent down to allow the lambs to feed when they are hungrie [*sic*]." (See Vegetable Lamb of Tartary.) This aspect is retained in the name for cotton in many European languages, such as German *Baumwolle*, which translates as "tree wool" (*Baum* means "tree"; *Wolle* means "wool"). By the end of the 16th century, cotton was cultivated throughout the warmer regions in Asia and the Americas.

## THE VEGETABLE LAMB OF TARTARY

India's cotton-processing sector gradually declined during British expansion in India and the establishment of colonial rule during the late 18th and early 19th centuries. This was largely due to the East India Company's de-industrialization of India, which forced the closing of cotton processing and manufacturing workshops in India, to ensure that Indian markets supplied only raw materials and were obliged to purchase manufactured textiles from Britain. The advent of the Industrial Revolution in Britain provided a great boost to cotton manufacture, as textiles emerged as Britain's leading export. In 1738 Lewis Paul and John Wyatt, of Birmingham, England, patented the Roller Spinning machine, and the flyer-and-bobbin system for drawing cotton to a more even thickness using two sets of rollers that traveled at different speeds.

Later, the invention of the spinning jenny in 1764 and Richard Arkwright's spinning frame (based on the Roller Spinning Machine) in 1769 enabled British weavers to produce cotton yarn and cloth at much higher rates. From the late eighteenth century onwards, the British city of Manchester acquired the nickname *"cottonopolis"* due to the cotton industry's omnipresence within the city, and Manchester's role as the heart of the global cotton trade. Production capacity was further improved by the invention of the cotton gin by Eli Whitney in 1793. Improving technology and increasing control of world markets allowed British traders to develop a commercial chain in which raw cotton fibers were (at first) purchased from colonial plantations, processed into cotton cloth in the mills of Lancashire, and then re-exported on British ships to captive colonial markets in West Africa, India, and China (via Shanghai and Hong Kong).

By the 1840s, India was no longer capable of supplying the vast quantities of cotton fibers needed by mechanised British factories, while shipping bulky, low-price cotton from India to Britain was time-consuming and expensive. This, coupled with the emergence of American cotton as a superior type (due to the longer, stronger fibers of the two domesticated native American species, *Gossypium hirsutum* and *Gossypium barbadense*), encouraged British traders to purchase cotton from plantations in the United States and the Caribbean. This was also much cheaper as it was produced by unpaid slaves. By the mid 19th century, "King Cotton" had become the backbone of the southern American economy. In the United States, cultivating and harvesting cotton became the leading occupation of slaves.

During the American Civil War, American cotton exports slumped due to a Union blockade on Southern ports, also because of a strategic decision by the Confederate Government to cut exports, hoping to force Britain to recognize the Confederacy or enter the war, prompting the main purchasers of cotton, Britain and France, to turn to Egyptian cotton. British and French traders invested heavily in cotton plantations and the Egyptian government of Viceroy Isma'il took out substantial loans from European bankers and stock exchanges. After the American Civil War ended in 1865, British and French traders abandoned Egyptian cotton and returned to cheap American exports, sending Egypt into a deficit spiral that led to the country declaring bankruptcy in 1876, a key factor behind Egypt's annexation by the British Empire in 1882.

## PICKING COTTON IN OKLAHOMA, USA, IN THE 1890S

During this time cotton cultivation in the British Empire, especially India, greatly increased to replace the lost production of the American South. Through tariffs and other restrictions the British government discouraged the production of cotton cloth in India; rather the raw fiber was sent to England for processing. The Indian patriot Mahatma Gandhi described the process:

> English people buy Indian cotton in the field, picked by Indian labor at seven cents a day, through an optional monopoly.

This cotton is shipped on British bottoms, a three-week journey across the Indian Ocean, down the Red Sea, across the Mediterranean, through Gibraltar, across the Bay of Biscay and the Atlantic Ocean to London. One hundred per cent profit on this freight is regarded as small.

The cotton is turned into cloth in Lancashire. You pay shilling wages instead of Indian pennies to your workers. The English worker not only has the advantage of better wages, but the steel companies of England get the profit of building the factories and machines. Wages; profits; all these are spent in England.

The finished product is sent back to India at European shipping rates, once again on British ships. The captains, officers, sailors of these ships, whose wages must be paid, are English. The only Indians who profit are a few lascars who do the dirty work on the boats for a few cents a day.

The cloth is finally sold back to the kings and landlords of India who got the money to buy this expensive cloth out of the poor peasants of India who worked at seven cents a day.

## PRISONERS FARMING COTTON UNDER TRUSTY SYSTEM - 1911

In the United States, Southern cotton provided capital for the continuing development of the North. The cotton produced by enslaved African Americans not only helped the South but also enriched Northern merchants. Much of the Southern cotton was transhipped through the northern ports. Cotton remained a key crop in the southern economy after emancipation and the end of the civil war in 1865. Across the South, sharecropping evolved, in which free black farmers worked on white-owned cotton plantations in return for a share of the profits. Cotton plantations required vast labor forces to hand-pick cotton, and it was not until the 1950s that reliable harvesting machinery was introduced into the South (prior to this, cotton-harvesting machinery had been too clumsy to pick cotton without shredding the fibers). During the early twentieth century, employment in the cotton industry fell as machines began to replace laborers, and as the South's rural labor force dwindled during the First and Second World Wars. Today, cotton remains a major export of the southern United States, and a majority of the world's annual cotton crop is of the long-staple American variety

## Tangüis Cotton

In 1901, Peru's cotton industry suffered because of a fungus plague caused by a plant disease known as "Cotton wilt" and " "Fusarium wilt" (*Fusarium vasinfectum*). The plant disease, which spread throughout Peru, entered the plant by its roots and worked it's way up the stem until the plant was completely dried up. Fermín Tangüis a Puerto Rican agriculturist who lived in Peru, studied some species of the plant that were affected by the disease to a lesser extent and experimented in germination with the seeds of various cotton plants. In 1911, after 10 years of experimenting and failures, Tangüis was able to develop a seed which produced a superior cotton plant resistant to the disease. The seeds produced a plant that had a 40% longer (between 29 mm and 33 mm) and thicker fiber that did not break easily and required little water. The Tangüis cotton, as it became known, is the variety which is preferred by the Peruvian national textile industry. It constituted 75 percent of all the Peruvian cotton production, both for domestic use and apparel exports. The Tangüis cotton crop was estimated at 225,000 bales that year.

## CULTIVATION

Successful cultivation of cotton requires a long frost-free period, plenty of sunshine, and a moderate rainfall, usually from 600 to 1200mm (24 to 48 inches). Soils usually need to be fairly heavy, although the level of nutrients does not need to be exceptional. In general, these conditions are met within the seasonally dry tropics and subtropics in the Northern and Southern hemispheres, but a large proportion of the cotton grown today is cultivated in areas with less rainfall that obtain the water from irrigation. Production of the crop for a given year usually starts soon after harvesting the preceding autumn. Planting time in spring in the Northern hemisphere varies from the beginning of February to the beginning of June. The area of the United States known as the South Plains is the largest contiguous cotton-growing region in the world. It is heavily dependent on irrigation water drawn from the Ogallala Aquifer.

Cotton is a thirsty crop, and as water resources get tighter around the world, economies that rely on it face difficulties and conflict, as well as potential environmental problems. For example, cotton has led to desertification in areas of Uzbekistan, where it is a major export. In the days of the Soviet Union, the Aral Sea was tapped for agricultural irrigation, largely of cotton, and now salination is widespread.

## GENETICALLY MODIFIED COTTON

Genetically modified (GM) cotton was developed to reduce the heavy reliance on pesticides. The bacterium *Bacillus thuringiensis* naturally produces a chemical harmful only to a small fraction of insects, most notably the larvae of moths and butterflies, beetles, and flies, and harmless to other forms of life. The gene coding for BT toxin has been inserted into cotton, causing cotton to produce this natural insecticide in its tissues. In many regions the main pests in commercial cotton are lepidopteran larvae, which are killed by the *Bt* protein in the transgenic cotton that they eat. This eliminates the need to use large amounts of broad-spectrum insecticides to kill lepidopteran pests (some of which have developed pyrethroid resistance). This spares natural insect predators in the farm ecology and further contributes to non-insecticide pest management.

*Bt* cotton is ineffective against many cotton pests, however, such as plant bugs, stink bugs, aphids, etc.; depending on circumstances it may still be desirable to use insecticides against these. Genetically modified cotton is widely used throughout the world. However, researchers have recently published the first documented case of in-field pest resistance to GM cotton. The International Service for the Acquisition of Agri-biotech Applications (ISAAA) said that, worldwide, GM cotton was planted on an area of 67,000 km$^2$ in 2002. This is 20% of the worldwide total area planted in cotton. The U.S. cotton crop was 73% GM in 2003. The initial introduction of GM cotton proved to be a commercial and ecological disaster in Australia - the yields were far lower than predicted, and the cotton plants were cross-pollinated with other varieties of cotton. However, the introduction of a second variety of GM cotton led to 15% of Australian cotton being GM in 2003. eighty per cent of the crop was genetically modified in 2004, when the original GM variety was banned. Cotton has also been genetically modified for resistance to glyphosate (marketed as Roundup in the USA), an inexpensive and highly effective but broad-spectrum herbicide, during its early growth. This allows glyphosate to be used for weed control during the early growing season.

The GM cotton acreage in India continues to grow at a rapid rate increasing from 50,000 hectares in 2002 to 3.8 million hectares in 2006. The total cotton area in India is about 9.0 million hectares (the largest in the world or, about 25% of world cotton area) so GM cotton is now grown on 42% of the cotton area. This makes India the country with the largest area of GM cotton in the world, surpassing China (3.5 million hectares in 2006). The major reasons for this increase is a combination of increased farm income ($225/ha) and a reduction in pesticide use to control the Cotton Bollworm. Cotton has gossypol, a toxin that makes it inedible. However, scientists have silenced the gene that produces the toxin, making it a potential food crop.

## ORGANIC COTTON

Organic cotton is cotton that is grown without insecticide or pesticide. Worldwide, cotton is a pesticide-intensive crop, using approximately 25% of the world's insecticides and 10% of the world's pesticides. According to the World Health Organisation (WHO), 20,000 deaths occur each year from pesticide poisoning in developing countries, many of these from cotton farming. Organic agriculture uses methods that are ecological, economical, and socially sustainable and denies the use of agrochemicals and artificial fertilizers. Instead, organic agriculture uses crop rotation, the growing of different crops than cotton in alternative years. The use of insecticides is prohibited; organic agriculture uses natural enemies to suppress harmful insects. The production of organic cotton is more expensive than the production of conventional cotton. Although toxic pollution from synthetic chemicals is eliminated, other pollution-like problems may remain, particularly run-off. Organic cotton is produced in organic agricultural systems that produce food and fiber according to clearly established standards. Organic agriculture prohibits the use of toxic and persistent chemical pesticides and fertilizers, as well as genetically modified organisms. It seeks to build biologically diverse agricultural systems, replenish and maintain soil fertility, and promote a healthy environment.

## Pests and Weeds

The cotton industry relies heavily on chemicals such as fertilizers and insecticides, although a very small number of farmers are moving toward an organic model of production and organic cotton products are now available for purchase at limited locations. These are popular for baby clothes and diapers. Under most definitions, organic products do not use genetic engineering. Historically, in North America, one of the most economically destructive pests in cotton production has been the boll weevil. Due to the US Department of Agriculture's highly successful Boll Weevil Eradication Programme (BWEP), this pest has been eliminated from cotton in most of the United States. This programme, along with the introduction of genetically engineered "Bt cotton" (which contains a bacteria gene that codes for a plant-produced protein that is toxic to a number of pests such as tobacco budworm, cotton bollworm, and pink bollworm), has allowed a reduction in the use of synthetic insecticides.

## Mechanized Harvesting

Most cotton in the United States, Europe, and Australia is harvested mechanically, either by a cotton picker, a machine that removes the cotton from the boll without damaging the cotton plant, or by a cotton stripper, which strips the entire boll off the plant. Cotton strippers are used in regions where it is too windy to grow picker varieties of cotton, and usually after application of a chemical defoliant or the natural defoliation that occurs after a freeze. Cotton is a perennial crop in the tropics and without defoliation or freezing, the plant will continue to grow. Cotton continues to be picked by hand in developing countries such as Uzbekistan.

## Competition from Synthetic Fibers

The era of manufactured fibers began with the development of Rayon in France in the 1890s. Rayon is derived from a natural cellulose and cannot be considered synthetic, but requires extensive processing in a manufacturing process and led the less expensive replacement of more naturally derived materials. A succession of new synthetic fibers were introduced by the chemicals industry in the following decades. Acetate in fiber form was developed in 1924. Nylon, the first fiber synthesized entirely from petrochemicals, was introduced as a sewing thread by DuPont in 1936, followed by Dupont's acrylic in 1944. Some garments were created from fabrics based on these fibers, such as women's hosiery from nylon, but it was not until the introduction of polyester into the fiber marketplace in the early 1950s that the market for cotton came under threat The rapid uptake of polyester garments in the 1960s caused economic hardship in cotton exporting economies, especially in Central American countries such as Nicaragua where cotton production had boomed tenfold between 1950 and 1965 with the advent of cheap chemical pesticides. Cotton production recovered in the 1970s, but crashed to pre-1960 levels in the early 1990s.

Beginning as a self-help programme in the mid-1960s, the Cotton Research and Promotion Programme was organized by U.S. cotton producers in response to cotton's steady decline in market share. At that time, producers voted to set up a per-bale assessment system to fund the programme, with built-in safeguards to protect their investments. With the passage of the Cotton Research and Promotion Act of 1966, the programme joined forces and began battling synthetic competitors and re-establishing markets for cotton. Today, the success of this programme has made cotton the best-selling fiber in the U.S. and one of the best-selling fibers in the world.

Administered by the Cotton Board and conducted by Cotton Incorporated, the Cotton Research and Promotion Programme works to greatly increase the demand for and profitability of cotton through various research and promotion activities. It is funded by U.S. cotton producers and importers.

## USES

Cotton is used to make a number of textile products. These include terrycloth, used to make highly absorbent bath towels and robes; denim, used to make blue jeans; chambray, popularly used in the manufacture of blue work shirts (from which we get the term "blue-collar"); and corduroy, seersucker, and cotton twill. Socks, underwear, and most T-shirts are made from cotton. Bed sheets often are made from cotton. Cotton also is used to make yarn used in crochet and knitting. Fabric also can be made from recycled or recovered cotton that otherwise would be thrown away during the spinning, weaving, or cutting process. While many fabrics are made completely of cotton, some materials blend cotton with other fibers, including rayon and synthetic fibers such as polyester. It can either be used in knitted or woven fabrics, as it can be blended with elastine to make a strechier thread for knitted fabrics, and things such as stretch jeans. In addition to the textile industry, cotton is used in fishnets, coffee filters, tents, gunpowder (see Nitrocellulose), cotton paper, and in bookbinding. The first Chinese (from China) paper was made of cotton fiber. Fire hoses were once made of cotton.

The cottonseed which remains after the cotton is ginned is used to produce cottonseed oil, which, after refining, can be consumed by humans like any other vegetable oil. The cottonseed meal that is left generally is fed to livestock. During the American slavery period, cotton root bark was used in a folk remedy as an abortifacient, that is, to provoke abortion. Cotton linters are fine, silky fibers which adhere to the seeds of the cotton plant after ginning. These curly fibers typically are less than 1/8 in (3 mm) long. The term also may apply to the longer textile fiber staple lint as well as the shorter fuzzy fibers from some upland species. Linters are traditionally used in the manufacture of paper and as a raw material in the manufacture of cellulose.

Shiny cotton is a processed version of the fiber that can be made into cloth resembling satin for shirts and suits. However, its hydrophobic property of not easily taking up water makes it unfit for the purpose of bath and dish towels. The term Egyptian cotton refers to the extra long staple cotton grown in Egypt and favored for the luxury and upmarket brands worldwide. During the U.S. Civil War, with heavy European investments, Egyptian-grown cotton became a major alternate source for British textile mills. Egyptian cotton is more durable and softer than American Pima cotton, which is why it is more expensive. Pima cotton is American cotton that is grown in the south western states of the U.S. In South Asia, cotton is widely used in mattresses, which are the most common type of mattress used in that region.

## THE INTERNATIONAL COTTON TRADE

The United States, with sales of $4.9 billion, and Africa, with sales of $2.1 billion, are the largest exporters of raw cotton. Total international trade is $12 billion. Africa's share of the cotton trade has doubled since 1980. Neither area has a significant domestic textile industry, textile manufacturing having moved to developing nations in Eastern and South Asia such as India and China. In Africa cotton is grown by numerous small holders. Dunavant Enterprises, based in Memphis, Tennessee, is the leading cotton broker in Africa with hundreds of purchasing agents. It operates cotton gins in Uganda, Mozambique, and Zambia. In Zambia it often offers loans for seed and expenses to the 180,000 small farmers who grow cotton for it, as well as advice on farming methods. Cargill also purchases cotton in Africa for export.

The 25,000 cotton growers in the United States are heavily subsidized at the rate of $2 billion per year. The future of these subsidies is uncertain and has led to anticipatory expansion of cotton brokers' operations in Africa. Dunavant expanded in Africa by buying out local operations. This is only possible in former British colonies and Mozambique; former French colonies continue to maintain tight monopolies, inherited from their former colonialist masters, on cotton purchases at low fixed prices.

The five leading exporters of cotton are:

1. the United States;
2. Uzbekistan;
3. India;

4. Brazil; and
5. Burkina Faso. The largest non-producing importers are Bangladesh, Indonesia, Thailand, Russia, and Taiwan.

In India, the states of Maharashtra (26.63%), Gujarat (17.96%) and Andhra Pradesh (13.75%) are the leading cotton producing states, these states have a predominantly tropical wet and dry climate. In the United States, the state of Texas leads in total production while the state of California has the highest yield per acre in the world

## FAIR TRADE

Cotton is an enormously important commodity throughout the world. However, many farmers in developing countries receive a low price for their produce, or find it difficult to compete with developed countries.

This has led to an international dispute:

> On 27 September 2002 Brazil requested consultations with the US regarding prohibited and actionable subsidies provided to US producers, users and/or exporters of upland cotton, as well as legislation, regulations, statutory instruments and amendments thereto providing such subsidies (including export credits), grants, and any other assistance to the US producers, users and exporters of upland cotton.

On 8 September 2004, the Panel Report recommended that the United States "withdraw" export credit guarantees and payments to domestic user and exporters, and "take appropriate steps to remove the adverse effects or withdraw" the mandatory price-contingent subsidy measures.

In addition to concerns over subsidies, the cotton industries of some countries are criticized for employing child labor and damaging workers' health by exposure to pesticides used in production. The international production and trade situation has led to 'fair trade' cotton clothing and footwear, joining a rapidly growing market for organic clothing, fair fashion or so-called 'ethical fashion'. The fair trade system was initiated in 2005 with producers from Cameroon, Mali and Senegal.

## Critical Temperatures

- Favorable travel temperature range - no lower limit: 25 °C (77 °F)
- Optimum travel temperature: 20 °C (68 °F)
- Glow temperature: 205 °C (401 °F)
- Fire point: 210 °C (410 °F)
- Autoignition temperature: 407 °C (765 °F)
- Autoignition temperature (for oily cotton): 120 °C (248°F)

Cotton dries out, becomes hard and brittle and loses all elasticity at temperatures above 25°C (77°F). Extended exposure to light causes similar problems.

A temperature range of 25 °C (77 °F) to 35 °C (95°F) is the optimal range for mold development. At temperatures below 0°C (32 °F), rotting of wet cotton stops. Damaged cotton is sometimes stored at these temperatures to prevent further deterioration.

## British Standard Cotton Yarn Measures

- 1 thread = 54 inches (about 137 cm)
- 1 skein or rap = 80 threads (120 yards or about 109 m)
- 1 hank = 7 skeins (840 yards or about 768 m)
- 1 spindle = 18 hanks (15,120 yards or about 13.826 km)

# 11

# Wheat

## INTRODUCTION

Wheat (*Triticum* sp.) is a worldwide cultivated grass from the Levant area of the Middle East. Globally, after maize, wheat is the second most produced food among the cereal crops; rice ranks third. Wheat grain is a staple food used to make flour for leavened, flat and steamed breads; cookies, cakes, breakfast cereal, pasta, juice, noodles and couscous and for fermentation to make beer, alcohol, vodka or biofuel. Wheat is planted to a limited extent as a forage crop for livestock, and the straw can be used as fodder for livestock or as a construction material for roofing thatch. Although wheat supplies much of the world's dietary protein and food supply, as many as one in every 100 to 200 people has Coeliac disease, a condition which results from an immune system response to a protein found in wheat: gluten (based on figures for the United States).

Wheat originated in Southwest Asia in the area known as the Fertile crescent. The genetic relationships between wild and domesticated populations of both einkorn and emmer wheat indicate that the most likely site of domestication is near Diyarbakir in Turkey. Wild wheats were domesticated as part of the origins of agriculture in the Fertile Crescent. Cultivation and repeated harvesting and sowing of the grains of wild grasses led to the domestication of wheat through selection of mutant forms with tough ears that remained intact during harvesting, larger grains, and a tendency for the spikelets to stay on the stalk until harvested. Because of the loss of seed dispersal mechanisms, domesticated wheats have limited capacity to propagate in the wild.

The exact timing of the first appearance of domesticated wheats is currently uncertain, but is either in the PPNA period (9800-8800 cal BC) or the early-mid PPNB (8800-7500 cal BC). Domesticated einkorn and emmer wheat has been identified at three PPNA sites in the northern Levant, Iraq ed-Dubb, Jericho and Tell Aswad, but both the dating and the domesticated status of these cereals is disputed. Domesticated wheats (and other Neolithic founder crops) are unambiguously present at early-mid PPNB sites in the northern Levant, such as Ain Ghazal, Abu Hureyra and Tell Aswad, and in southeast Turkey at Cafer Höyük and Çayönü As a round figure, it is correct to say that wheats have been domesticated for about 10,000 years.

The cultivation of wheat began to spread beyond the Fertile Crescent during the Neolithic period, reaching the Aegean by 8500 cal BC and the Indian subcontinent by 6000 cal BC. By 5,000 years ago, wheat had reached Ethiopia, Great Britain, Ireland and Spain. A millennium later it reached China. Claims have been made for independent domestication of wheat outside the fertile crescent, but these lack evidence of the presence of wild wheats or of early domesticated wheat. Three thousand years ago agricultural cultivation with horse-drawn plows increased cereal grain production, as did the use of seed drills to replace broadcast sowing in the 18th century. Yields of wheat continued to increase, as new land came under cultivation and with improved agricultural husbandry involving the use of fertilizers, threshing machines and reaping machines, tractor-drawn cultivators and planters, and varieties adapted to intensive cultivation (see green revolution and Norin 10 wheat).

## Genetics

Wheat genetics is more complicated than that of most other domesticated species. Some wheat species are diploid, with two sets of chromosomes, but many are stable polyploids, with four sets of chromosomes (tetraploid) or six (hexaploid).

- Einkorn wheat (*T. monococcum*) is diploid.
- Most tetraploid wheats (e.g. emmer and durum wheat) are derived from wild emmer, *T. dicoccoides*. Wild emmer is the result of a hybridization between two diploid wild grasses, *T. urartu* and a wild goatgrass such as *Aegilops searsii* or *Ae. speltoides*. The hybridization that formed wild emmer occurred in the wild, long before domestication.
- Hexaploid wheats evolved in farmers' fields. Either domesticated emmer or durum wheat hybridized with yet another wild diploid grass (*Aegilops tauschii*) to make the hexaploid wheats, spelt wheat and bread wheat.

## PLANT BREEDING

### Wheat

In traditional agricultural systems wheat populations often consist of landraces, informal farmer-maintained populations that often maintain high levels of morphological diversity. Although landraces of wheat are no longer grown in Europe and North America, they continue to be important elsewhere. The origins of formal wheat breeding lie in the nineteenth century, when single line varieties were created through selection of seed from a single plant noted to have desired properties. Modern wheat breeding developed in the first years of the twentieth century and was closely linked to the development of Mendelian genetics. The standard method of breeding inbred wheat cultivars is by crossing two lines using hand emasculation, then selfing or inbreeding the progeny. Selections are *identified* (shown to have the genes responsible for the varietal differences) ten or more generations before release as a variety or cultivar.

F1 hybrid wheat cultivars should not be confused with wheat cultivars deriving from standard plant breeding. Heterosis or hybrid vigor (as in the familiar F1 hybrids of maize) occurs in common (hexaploid) wheat, but it is difficult to produce seed of hybrid cultivars on a commercial scale as is done with maize because wheat flowers are complete and normally self-pollinate Commercial hybrid wheat seed has been produced using chemical hybridizing agents, plant growth regulators that selectively interfere with pollen development, or naturally occurring cytoplasmic male sterility systems. Hybrid wheat has been a limited commercial success in Europe (particularly France), the USA and South Africa.

The major breeding objectives include high grain yield, good quality, disease and insect resistance and tolerance to abiotic stresses include mineral, moisture and heat tolerance. The major diseases in temperate environments include Fusarium head blight, leaf rust and stem rust, whereas in tropical areas spot blotch (wheat) (also known as Helminthosporium leaf blight). See physiological and molecular wheat breeding.

The four wild species of wheat, along with the domesticated varieties einkorn, emmer and spelt, have hulls (in German, *Spelzweizen*). This more primitive morphology consists of toughened glumes that tightly enclose the grains, and (in domesticated wheats) a semi-brittle rachis that breaks easily on threshing. The result is that when threshed, the wheat ear breaks up into spikelets. To obtain the grain, further processing, such as milling or pounding, is needed to remove the hulls or husks. In contrast, in free-threshing (or naked) forms such as durum wheat and common wheat, the glumes are fragile and the rachis tough. On threshing, the chaff breaks up, releasing the grains. Hulled wheats are often stored as spikelets because the toughened glumes give good protection against pests of stored grain.

## SACK OF WHEAT

There are many botanical classification systems used for wheat species, discussed in a separate article on Wheat taxonomy. The name of a wheat species from one information source may not be the name of a wheat species in another. Within a species, wheat cultivars are further classified by wheat breeders and farmers in terms of growing season, such as winter wheat vs. spring wheat, by gluten content, such as hard wheat (high protein content) vs. soft wheat (high starch content), or by grain color (red, white or amber). In British English wheat may be referred to as corn.

### Major cultivated species of wheat

- *Common wheat or Bread wheat* — (*T. aestivum*) A hexaploid species that is the most widely cultivated in the world.

- *Durum* — (*T. durum*) The only tetraploid form of wheat widely used today, and the second most widely cultivated wheat.
- *Einkorn* — (*T. monococcum*) A diploid species with wild and cultivated variants. Domesticated at the same time as emmer wheat, but never reached the same importance.
- *Emmer* — (*T. dicoccon*) A tetraploid species, cultivated in ancient times but no longer in widespread use.
- *Spelt* — (*T. spelta*) Another hexaploid species cultivated in limited quantities.

## In the United States

Classes used in the United States are:

- *Durum* — Very hard, translucent, light colored grain used to make semolina flour for pasta.
- *Hard Red Spring* — Hard, brownish, high protein wheat used for bread and hard baked goods. Bread Flour and high gluten flours are commonly made from hard red spring wheat. It is primarily traded at the Minneapolis Grain Exchange.
- *Hard Red Winter* — Hard, brownish, mellow high protein wheat used for bread, hard baked goods and as an adjunct in other flours to increase protein in pastry flour for pie crusts. Some brands of unbleached all-purpose flours are commonly made from hard red winter wheat alone. It is primarily traded by the Kansas City Board of Trade. One variety is known as "turkey red wheat", and was brought to Kansas by Mennonite immigrants from Russia.
- *Soft Red Winter* — Soft, low protein wheat used for cakes, pie crusts, biscuits, and muffins. Cake flour, pastry flour, and some self-rising flours with baking powder and salt added for example, are made from soft red winter wheat. It is primarily traded by the Chicago Board of Trade.
- *Hard White* — Hard, light colored, opaque, chalky, medium protein wheat planted in dry, temperate areas. Used for bread and brewing.

- *Soft White* — Soft, light colored, very low protein wheat grown in temperate moist areas. Used for pie crusts and pastry. Pastry flour, for example, is sometimes made from soft white winter wheat.

Hard wheats are harder to process and red wheats may need bleaching. Therefore, soft and white wheats usually command higher prices than hard and red wheats on the commodities market.

## As a Food

Raw wheat can be powdered into flour, germinated and dried creating malt, crushed and de-branned into cracked wheat, parboiled (or steamed), dried, crushed and de-branned into bulgur, or processed into semolina, pasta, or roux. They are a major ingredient in such foods as bread, breakfast cereals (e.g. Wheatena, Cream of Wheat, Shredded Wheat), porridge, crackers, biscuits, Muesli, pancakes, cakes, gravy and boza (a fermented beverage).

## Nutrition

100 grams of hard red winter wheat contain about 12.6 grams of protein, 1.5 grams of total fat, 71 grams of carbohydrate (by difference), 12.2 grams of dietary fiber, and 3.2 mg of iron (17% of the daily requirement); the same weight of hard red spring wheat contains about 15.4 grams of protein, 1.9 grams of total fat, 68 grams of carbohydrate (by difference), 12.2 grams of dietary fiber, and 3.6 mg of iron (20% of the daily requirement).

Gluten, a protein found in wheat (and other Triticeae), cannot be tolerated by people with celiac disease (an autoimmune disorder in ~1% of Indo-European populations.

Much of the carbohydrate fraction of wheat is starch. Wheat starch is an important commercial product of wheat, but second in economic value to wheat gluten. The principal parts of wheat flour are gluten and starch. These can be separated in a kind of home experiment, by mixing flour and water to form a small ball of dough, and kneading it gently while rinsing it in a bowl of water. The starch falls out of the dough and sinks to the bottom of the bowl, leaving behind a ball of gluten.

## Health Concerns

Roughly 1% of the population has coeliac or celiac disease—a condition that is caused by an adverse immune system reaction to gliadin, a gluten protein found in wheat (and similar proteins of the tribe Triticeae which includes other cultivars such as barley and rye). Upon exposure to gliadin, the enzyme tissue transglutaminase modifies the protein, and the immune system cross-reacts with the bowel tissue, causing an inflammatory reaction. That leads to flattening of the lining of the small intestine, which interferes with the absorption of nutrients.

The only effective treatment is a lifelong gluten-free diet. While the disease is caused by a reaction to wheat proteins, it is not the same as wheat allergy. Harvested wheat grain that enters trade is classified according to grain properties for the purposes of the commodities market.

Wheat buyers use the classifications to help determine which wheat to purchase as each class has special uses. Wheat producers determine which classes of wheat are the most profitable to cultivate with this system. Wheat is widely cultivated as a cash crop because it produces a good yield per unit area, grows well in a temperate climate even with a moderately short growing season, and yields a versatile, high-quality flour that is widely used in baking. Most breads are made with wheat flour, including many breads named for the other grains they contain like most rye and oat breads. The popularity of foods made from wheat flour creates a large demand for the grain, even in economies with significant food surpluses.

In 2007 there was a dramatic rise in the price of wheat due to freezes and flooding in the northern hemisphere and a drought in Australia. Wheat futures in September, 2007 for December and March delivery had risen above $9.00 a bushel, prices never seen before there were complaints in Italy about the high price of pasta. This followed a wider trend of escalating food prices around the globe, driven in part by climatic conditions such as drought in Australia, the diversion of arable land to other uses (such as producing government-subsidised bio-oil crops), and later by some food-producing nations placing bans or restrictions on exports in order to satisfy their own consumers.

Other drivers affecting wheat prices include the movement to bio fuels (in 2008, a third of corn crops in the US are expected to be devoted to ethanol production) and rising incomes in developing countries, which is causing a shift in eating patterns from predominantly rice to more meat based diets (a rise in meat production equals a rise in grain consumption - seven kilograms of grain is required to produce one kilogram of beef.

## PRODUCTION AND CONSUMPTION STATISTICS

In 2003, global per capita wheat consumption was 67 kg, with the highest per capita consumption (239 kg) found in Kyrgyzstan.

Unlike rice, wheat production is more widespread globally though China's share is almost one-sixth of the world.

## AGRONOMY

While winter wheat lies dormant during a winter freeze, wheat normally requires between 110 and 130 days between planting and harvest, depending upon climate, seed type, and soil conditions. Crop management decisions require the knowledge of stage of development of the crop. In particular, spring fertilizer applications, herbicides, fungicides, growth regulators are typically applied at specific stages of plant development.

For example, current recommendations often indicate the second application of nitrogen be done when the ear (not visible at this stage) is about 1 cm in size (Z31 on Zadoks scale). Knowledge of stages is also interesting to identify periods of higher risk, in terms of climate. For example, the meiosis stage is extremely susceptible to low temperatures (under 4 °C) or high temperatures (over 25 °C). Farmers also benefit from knowing when the flag leaf (last leaf) appears as this leaf represents about 75% of photosynthesis reactions during the grain-filling period and as such should be preserved from disease or insect attacks to ensure a good yield. Several systems exist to identify crop stages, with the Feekes and Zadoks scales being the most widely used. Each scale is a standard system which describes successive stages reached by the crop during the agricultural season.

# 12

# Rice

## INTRODUCTION

Rice is a cereal foodstuff which forms an important part of the diet of many people worldwide and as such it is a staple food for many. Domesticated rice comprises two species of food crops in the *Oryza* genus of the Poaceae ("true grass") family: Asian rice, *Oryza sativa* is native to tropical and subtropical southern Asia; African rice, *Oryza glaberrima,* is native to West Africa. The name wild rice is usually used for species of the different but related genus *Zizania,* both wild and domesticated, although the term may be used for primitive or uncultivated varieties of *Oryza.*

Rice is grown as a monocarpic annual plant, although in tropical areas it can survive as a perennial and can produce a ratoon crop and survive for up to 20 years. Rice can grow to 1–1.8 m tall, occasionally more depending on the variety and soil fertility. The grass has long, slender leaves 50–100 cm long and 2–2.5 cm broad. The small wind-pollinated flowers are produced in a branched arching to pendulous inflorescence 30–50 cm long.

The edible seed is a grain (caryopsis) 5–12 mm long and 2–3 mm thick. Rice is a staple food for a large part of the world's human population, especially in tropical Latin America, and East, South and Southeast Asia, making it the second-most consumed cereal grain. A traditional food plant in Africa, Rice has the potential to improve nutrition, boost food security, foster rural development and support sustainable landcare. Rice provides more than one fifth of the calories consumed worldwide by humans. In early 2008, some governments and retailers began rationing supplies of the grain due to fears of a global rice shortage.

Rice cultivation is well-suited to countries and regions with low labor costs and high rainfall, as it is very labour-intensive to cultivate and requires plenty of water for cultivation. On the other hand, mechanized cultivation is extremely oil-intensive, more than other food products with the exception of beef and dairy products. Rice can be grown practically anywhere, even on a steep hill or mountain. Although its species are native to South Asia and certain parts of Africa, centuries of trade and exportation have made it commonplace in many cultures.

The traditional method for cultivating rice is flooding the fields whilst, or after, setting the young seedlings. This simple method requires sound planning and servicing of the water damming and channeling, but reduces the growth of less robust weed and pest plants that have no submerged growth state, and deters vermin. While with rice growing and cultivation the flooding is not mandatory, all other methods of irrigation require higher effort in weed and pest control during growth periods and a different approach for fertilizing the soil.

## Classification

There are two species of domesticated rice, *Oryza sativa* (Asian) and *Oryza glaberrima* (African). *Oryza sativa* contains two major subspecies: the sticky, short-grained *japonica* or *sinica* variety, and the non-sticky, long-grained *indica* variety. *Japonica* are usually cultivated in dry fields, in temperate East Asia, upland areas of Southeast Asia and high elevations in South Asia, while *indica* are mainly lowland rices, grown mostly submerged, throughout tropical Asia. A third subspecies, which is broad-grained and thrives under tropical conditions, was identified based on morphology and initially called *javanica*, but is now known as *tropical japonica*. Examples of this variety include the medium grain "Tinawon" and "Unoy" cultivars, which are grown in the high-elevation rice terraces of the Cordillera Mountains of northern Luzon, Philippines.

Isozymes to sort *Oryza sativa* into six groups: *japonica, aromatic, indica, aus, rayada,* and *ashina*. The SSRs to sort *Oryza sativa* into five groups; *temperate japonica, tropical japonica* and *aromatic* comprise the *japonica* varieties, while *indica* and *aus* comprise the *indica* varieties.

## Etymology

According to the *Microsoft Encarta Dictionary* (2004) and the *Chambers Dictionary of Etymology* (1988), the word 'rice' has an Indo-Iranian origin. It came to English from Greek *óryza,* via Latin *oriza,* Italian *riso* and finally Old French *ris* (the same as present day French *riz*). It has been speculated that the Indo-Iranian *vrihi* itself is borrowed from a Dravidian vari (< PDr. *warinci or even a Munda language term for rice, or the Tamil name *arisi* from which the Arabic *ar-ruzz,* from which the Portuguese and Spanish word *arroz* originated.

## PREPARATION AS FOOD

The seeds of the rice plant are first milled using a rice huller to remove the chaff (the outer husks of the grain). At this point in the process, the product is called brown rice. The milling may be continued, removing the 'bran' (i.e. the rest of the husk and the germ), thereby creating white rice. White rice, which keeps longer, lacks some important nutrients; in a limited diet which does not supplement the rice, brown rice helps to prevent the deficiency disease beriberi.

White rice may be also buffed with glucose or talc powder (often called *polished rice,* though this term may also refer to white rice in general), parboiled, or processed into flour. White rice may also be enriched by adding nutrients, especially those lost during the milling process. While the cheapest method of enriching involves adding a powdered blend of nutrients that will easily wash off (in the United States, rice which has been so treated requires a label warning against rinsing), more sophisticated methods apply nutrients directly to the grain, coating the grain with a water insoluble substance which is resistant to washing.

Despite the hypothetical health risks of talc (such as stomach cancer) talc-coated rice remains the norm in some countries due to its attractive shiny appearance, but it has been banned in some and is no longer widely used in others such as the United States. Even where talc is not used, glucose, starch, or other coatings may be used to improve the appearance of the grains; for this reason, many rice lovers still recommend washing all rice in order to create a better-tasting rice with a better consistency, despite the recommendation of suppliers. Much of the rice produced today is water polished.

Rice bran, called *nuka* in Japan, is a valuable commodity in Asia and is used for many daily needs. It is a moist, oily inner layer which is heated to produce an oil. It is also used as a pickling bed in making rice bran pickles and Takuan.

Raw rice may be ground into flour for many uses, including making many kinds of beverages such as amazake, horchata, rice milk, and sake. Rice flour does not contain gluten and is suitable for people on a gluten-free diet. Rice may also be made into various types of noodles. Raw wild or brown rice may also be consumed by raw-foodist or fruitarians if soaked and sprouted (usually 1 week to 30 days).

Processed rice seeds must be boiled or steamed before eating. Cooked rice may be further fried in oil or butter, or beaten in a tub to make mochi. Rice is a good source of protein and a staple food in many parts of the world, but it is not a complete protein: it does not contain all of the essential amino acids in sufficient amounts for good health, and should be combined with other sources of protein, such as nuts, seeds, beans or meat.

Rice, like other cereal grains, can be puffed (or popped). This process takes advantage of the grains' water content and typically involves heating grains in a special chamber. Further puffing is sometimes accomplished by processing pre-puffed pellets in a low-pressure chamber. The ideal gas law means that either lowering the local pressure or raising the water temperature results in an increase in volume prior to water evaporation, resulting in a puffy texture. Bulk raw rice density is about 0.9 g/cm$^3$. It decreases more than tenfold when puffed.

## Cooking

There are many varieties of rice; for many purposes the main distinction is between long- and medium-grain rice. The grains of long-grain rice tend to remain intact after cooking; medium-grain rice becomes more sticky. Medium-grain rice is used for sweet dishes, and for risotto and many Spanish dishes.

Rice is cooked by boiling or steaming, and absorbs water during cooking. It can be cooked in just as much water as it absorbs (the absorption method), or in a large quantity of water which is drained before serving (the rapid-boil method). Electric rice cookers, popular in Asia and Latin America, simplify the process of cooking rice. Rice is often heated in oil before boiling, or oil is added to the water; this is thought to make the cooked rice less sticky.

In Arab cuisine rice is an ingredient of many soups and dishes with fish, poultry, and other types of meat. It is also used to stuff vegetables or is wrapped in grape leaves. When combined with milk, sugar and honey, it is used to make desserts. In some regions, such as Tabaristan, bread is made using rice flour. Medieval Islamic texts spoke of medical uses for the plant.

Rice may also be made into rice porridge (also called congee or rice gruel) by adding more water than usual, so that the cooked rice is saturated with water to the point that it becomes very soft, expanded, and fluffy. Rice porridge is commonly eaten as a breakfast food, and is also a traditional food for the sick.

Rice may be soaked prior to cooking, which decreases cooking time. For some varieties, soaking improves the texture of the cooked rice by increasing expansion of the grains. In some countries parboiled rice, also known as Minute rice or easy-cook rice, is popular. Parboiled rice is subjected to a steaming or parboiling process while still a brown rice. This causes nutrients from the outer husk to move into the grain itself. The parboil process causes a gelatinisation of the starch in the grains. The grains become less brittle, and the colour of the milled grain changes from white to yellow. The rice is then dried, and can then be milled as usual or used as brown rice. Milled parboiled rice is nutritionally superior to standard milled rice. Parboiled rice has an additional benefit in that it does not stick to the pan during cooking as happens when cooking regular white rice.

A nutritionally superior method of preparing brown rice known as GABA Rice or GBR (Germinated Brown Rice) may be used. This involves soaking washed brown rice for 20 hours in warm water (38°C or 100°F) prior to cooking it. This process stimulates germination, which activates various enzymes in the rice. By this method, a result of research carried out for the United Nations Year of Rice, it is possible to obtain a more complete amino acid profile, including GABA.

Cooked rice can contain Bacillus cereus spores which produce an emetic toxin when left at 4–60°C . When storing cooked rice for use the next day, rapid cooling is advised to reduce the risk of contamination.

## RICE GROWING ECOLOGY

— Rice can be grown in different ecologies, depending upon water availability.

— Lowland, rainfed, which is drought prone, favors medium depth; waterlogged, submergence, and flood prone.

— Lowland, irrigated, grown in both the wet season and the dry season.

— Deep water or floating rice.

### Coastal Wetland

Upland rice, also known as 'Ghaiya rice', well known for its drought tolerance.

### Cultivation

Based on one chloroplast and two nuclear gene regions, Londo *et al* (2006) conclude that rice was domesticated at least twice—*indica* in eastern India, Myanmar and Thailand; and *japonica* in southern China—though they concede that there is archaeological and genetic evidence for a single domestication of rice in the lowlands of China.

Because the functional allele for non-shattering—the critical indicator of domestication in grains—as well as five other single nucleotide polymorphisms, is identical in both *indica* and *japonica*, Vaughan *et al* (2008) determined that there was a single domestication event for *Oryza sativa* in the region of the Yangtze river valley.

## CONTINENTAL EAST ASIA

Rice appears to have been used by the Early Neolithic populations of Lijiacun and Yunchanyan. Rice cultivation began in China ca. 11,500 BP. Bruce Smith, an archaeologist at the Smithsonian Institution in Washington, D.C., who has written on the origins of agriculture, says that evidence has been mounting that the Yangtze was probably the site of the earliest rice cultivation.

The collection of wild rice in the Late Pleistocene had, by 6400 BC, led to the use of primarily domesticated rice. Morphological studies of rice phytoliths from the Diaotonghuan archaeological site clearly show the transition from the collection of wild rice to the cultivation of domesticated rice. The large number of wild rice phytoliths at the Diaotonghuan level dating from 12,000-11,000 BP indicates that wild rice collection was part of the local means of subsistence. Changes in the morphology of Diaotonghuan phytoliths dating from 10,000-8,000 BP show that rice had by this time been domesticated. Analysis of Chinese rice residues from Pengtoushan which were C14 dated to 8200-7800 BCE show that rice had been domesticated by this time.

In 1998, it is reported that the earliest of 14 AMS or radiocarbon dates on rice from at least nine Early to Middle Neolithic sides is no older than 7000 BC, that rice from the Hemudu and Luojiajiao sites indicates that rice domestication likely began before 5000 BC, but that most sites in China from which rice remains have been recovered are younger than 5000 BC.

## South Asia

Wild *Oryza* rice appeared in the Belan and Ganges valley regions of northern India as early as 4530 BC and 5440 BC respectively although many believe it may have appeared earlier. The Encyclopedia Britannica—on the subject of the first certain cultivated rice—holds.

## Korean Peninsula and Japanese Archipelago

Mainstream archaeological evidence derived from palaeoethnobotanical investigations indicate that dry-land rice was introduced to Korea and Japan some time between 3500 and 1200 BC. The cultivation of rice in Korea and Japan during that time occurred on a small-scale, fields were impermanent plots, and evidence shows that in some cases domesticated and wild grains were planted together. The technological, subsistence, and social impact of rice and grain cultivation is not evident in archaeological data until after 1500 BC. For example, intensive wet-paddy rice agriculture was introduced into Korea shortly before or during the Middle Mumun Pottery Period (c. 850–550 BC) and reached Japan by the Final Jomon or Initial Yayoi circa 300 BC.

In 2003, Korean archaeologists alleged that they discovered burnt grains of domesticated rice in Soro-ri, Korea, which dated to 13,000 BC. These predate the oldest grains in China, which were dated to 10,000 BC, and potentially challenge the mainstream explanation that domesticated rice originated in China The findings were received by academia with strong skepticism, and the results and their publicizing has been cited as being driven by a combination of nationalist and regional interests.

## Southeast Asia

Using water buffalo to plough rice fields in Java; Indonesia is the world's third largest paddy rice producer and its cultivation has transformed much of the country's landscape. Rice is the staple for all classes in contemporary South East Asia, from Myanmar to Indonesia. In Indonesia, evidence of wild Oryza rice on the island of Sulawesi dates from 3000 BCE. The evidence for the earliest cultivation, however, comes from eighth century stone inscriptions from Java, which show kings levied taxes in rice. Divisions of labour between men, women, and animals that are still in place in Indonesian rice cultivation, can be seen carved into the ninth-century Prambanan temples in Central Java. In the sixteenth century, Europeans visiting the Indonesian islands saw rice as a new prestige food served to the aristocracy during ceremonies and feasts. Rice production in Indonesian history is linked to the development of iron tools and the domestication of water buffalo for cultivation of fields and manure for fertilizer. Once covered in dense forest, much of the Indonesian landscape has been gradually cleared for permanent fields and settlements as rice cultivation developed over the last fifteen hundred years.

In the Philippines, the greatest evidence of rice cultivation since ancient times can be found in the Cordillera Mountain Range of Luzon in the provinces of Apayao, Benguet, Mountain Province and Ifugao. The Banaue Rice Terraces (Tagalog: Hagdan-hagdang Palayan ng Banaue) are 2000 to 3000-year old terraces that were carved into the mountains by ancestors of the Batad indigenous people. It is commonly thought that the terraces were built with minimal equipment, largely by hand. The terraces are located approximately 1,500 meters (5000 ft) above sea level and cover 10,360 square kilometers (about 4,000 square miles) of mountainside. They are fed by an ancient irrigation system from the rainforests above the terraces. It is said that if the steps are put end to end it would encircle half the globe. The Rice Terraces (a UNESCO World Heritage Site) are commonly referred to by Filipinos as the "Eighth Wonder of the World".

Evidence of wet rice cultivation as early as 2200 BC has been discovered at both Ban Chiang and Ban Prasat in Thailand. By the 19th Century, encroaching European expansionism in the area increased rice production in much of South East Asia, and Thailand, then known as Siam. British Burma became the world's largest exporter of rice, from the turn of the 20th century up till the 1970s, when neighbouring Thailand exceeded Burma.

## AFRICA

African rice has been cultivated for 3500 years. Between 1500 and 800 BC, *O. glaberrima* propagated from its original centre, the Niger River delta, and extended to Senegal. However, it never developed far from its original region. Its cultivation even declined in favour of the Asian species, possibly brought to the

African continent by Arabs coming from the east coast between the 7th and 11th centuries CE. In parts of Africa under Islam, rice was chiefly grown in southern Morocco. During the tenth century rice was also brought to east Africa by Muslim traders. Although, the diffusion of rice in much sub-Saharan Africa remains uncertain, Muslims brought it to the region stretching from Lake Chad to the White Nile. The actual and hypothesized cultivation of rice (areas shown in green) in the Old World (both Muslim and non-Muslim regions) during Islamic times (700-1500). Cultivation of rice during pre-Islamic times have been shown in orange.

## Middle East

*O. sativa* was introduced to the Middle East in Hellenistic times, and was familiar to both Greek and Roman writers. They report that a large sample of rice grains was recovered from a grave at Susa in Iran (dated to the first century AD) at one end of the ancient world, while at the same time rice was grown in the Po valley in Italy. However, Pliny the Elder writes that rice (*oryza*) is grown only in "Egypt, Syria, Cilicia, Asia Minor and Greece".

After the rise of Islam, rice was grown anywhere there was enough water to irrigate it. Thus, desert oases, river valleys, and swamp lands were all important sources of rice during the Muslim Agricultural Revolution.

In Iraq rice was grown in some areas of southern Iraq. With the rise of Islam it moved north to Nisibin, the southern shores of the Caspian Sea and then beyond the Muslim world into the valley of Volga. In Israel, rice came to be grown in the Jordan valley. Rice is also grown in Yemen.

## Europe

The Muslims (later known as Moors) brought Asiatic rice to the Iberian Peninsula in the tenth century. Records indicate it was grown in Valencia and Majorca. In Majorca, rice cultivation seems to have stopped after the Christian conquest, although historians are not certain. Muslims also brought rice to Sicily, where it was an important crop. After the middle of the 15th century, rice spread throughout Italy and then France, later propagating to all the continents during the age of European exploration.

## United States

In 1694, rice arrived in South Carolina, probably originating from Madagascar. In the United States, colonial South Carolina and Georgia grew and amassed great wealth from the slave labor obtained from the Senegambia area of West Africa and from coastal Sierra Leone. At the port of Charleston, through which 40% of all American slave imports passed, slaves from this region of Africa brought the highest prices, in recognition of their prior knowledge of rice culture, which was put to use on the many rice plantations around Georgetown, Charleston, and Savannah. From the slaves, plantation owners learned how to dyke the marshes and periodically flood the fields. At first the rice was milled by hand with wooden paddles, then winnowed in sweetgrass baskets (the making of which was another skill brought by the slaves). The invention of the rice mill increased profitability of the crop, and the addition of water power for the mills in 1787 by millwright Jonathan Lucas was another step forward. Rice culture in the southeastern U.S. became less profitable with the loss of slave labor after the American Civil War, and it finally died out just after the turn of the 20th century. Today, people can visit the only remaining rice plantation in South Carolina that still has the original winnowing barn and rice mill from the mid-1800s at the historic Mansfield Plantation in Georgetown, SC. The predominant strain of rice in the Carolinas was from Africa and was known as "Carolina Gold." The cultivar has been preserved and there are current attempts to reintroduce it as a commercially grown crop.

## AMERICAN LONG-GRAIN RICE

In the southern United States, rice has been grown in southern Arkansas, Louisiana, and east Texas since the mid 1800s. Many Cajun farmers grew rice in wet marshes and low lying prairies. In recent years rice production has risen in North America, especially in the Mississippi River Delta areas in the states of Arkansas and Mississippi.

Rice cultivation began in California during the California Gold Rush, when an estimated 40,000 Chinese labourers immigrated to the state and grew small amounts of the grain for their own consumption. However, commercial production began only in 1912 in the town of Richvale in Butte County. By 2006, California produced the second largest rice crop in the United States after Arkansas, with production concentrated in six counties north of Sacramento. Unlike the Mississippi Delta region, California's production is dominated by short- and medium-grain *japonica* varieties, including cultivars developed for the local climate such as Calrose, which makes up as much as eighty five percent of the state's crop.

References to wild rice in the Americas are to the unrelated *Zizania palustris*

More than 100 varieties of rice are commercially produced primarily in six states (Arkansas, Texas, Louisiana, Mississippi, Missouri, and California) in the U.S. According to estimates for the 2006 crop year, rice production in the U.S. is valued at $1.88 billion, approximately half of which is expected to be exported. The U.S. provides about 12% of world rice trade The majority of domestic utilization of U.S. rice is direct food use (58%), while 16 per cent is used in processed foods and beer respectively. The remaining 10 percent is found in pet food.

Although attempts to grow rice in the well-watered north of Australia have been made for many years, they have consistently failed because of inherent iron and manganese toxicities in the soils and destruction by pests.

In the 1920s it was seen as a possible irrigation crop on soils within the Murray-Darling Basin that were too heavy for the cultivation of fruit and too infertile for wheat.

Because irrigation water, despite the extremely low runoff of temperate Australia, was (and remains) very cheap, the growing of rice was taken up by agricultural groups over the following decades. Californian varieties of rice were found suitable for the climate in the Riverina, and the first mill opened at Leeton in 1951.

Even before this Australia's rice production greatly exceeded local needs and rice exports to Japan have become a major source of foreign currency. Above-average rainfall from the 1950s to the middle 1990s encouraged the expansion of the Riverina rice industry, but its prodigious water use in a practically waterless region began to attract the attention of environmental scientists. These became severely concerned with declining flow in the Snowy River and the lower Murray river.

Although rice growing in Australia is exceedingly efficient and highly profitable due to the cheapness of land, several recent years of severe drought have led many to call for its elimination because of its effects on extremely fragile aquatic ecosystems. Politicians, however, have not made any plan to reduce rice growing in southern Australia.

## WORLD PRODUCTION AND TRADE

World production of rice has risen steadily from about 200 million tonnes of paddy rice in 1960 to 600 million tonnes in 2004. Milled rice is about 68% of paddy rice by weight. In the year 2004, the top three producers were China (26% of world production), India (20%), and Indonesia (9%).

World trade figures are very different, as only about 5-6% of rice produced is traded internationally. The largest three exporting countries are Thailand (26% of world exports), Vietnam (15%), and the United States (11%), while the largest three importers are Indonesia (14%), Bangladesh (4%), and Brazil (3%). Although China and India are the top two largest producers of rice in the world, both of countries consume the majority of the rice produced domestically leaving little to be traded internationally.

### Price

In March to May 2008, the price of rice rose greatly due to a rice shortage. In late April 2008, rice prices hit 24 cents a pound, twice the price that it was seven months earlier. On the 30th of April, 2008, Thailand announced the project of the creation of the Organisation of Rice Exporting Countries (OREC) with the potential to develop into a price-fixing cartel for rice.

### Worldwide Consumption

Between 1961 and 2002, per capita consumption of rice increased by 40%. Rice consumption is highest in Asia, where average per capita consumption is higher than 80 kg/person per year. In the subtropics such as South America, Africa, and the Middle East, per capita consumption averages between 30 and 60 kg/person per year. People in the developed West, including Europe and the United States, consume less than 10 kg/person per year. Rice is the most important crop in Asia. In Cambodia, for example, 90% of the total agricultural area is used for rice production. See *The Burning of the Rice* by Don Puckridge for the story of rice production in Cambodia. U.S. rice consumption has risen sharply over the past 25 years, fueled in part by commercial applications such as beer production Almost one in five adult Americans now report eating at least half a serving of white or brown rice per day.

## Environmental Impacts

In many countries where rice is the main cereal crop, rice cultivation is responsible for most of the methane emissions. Farmers in some of the arid regions try to cultivate rice using groundwater bored through pumps, thus increasing the chances of famine in the long run Rice also requires much more water to produce than other grains.

As sea levels rise, rice will become more inclined to remain flooded for longer periods of time. Longer stays in water cuts the soil off from atmospheric oxygen and causes fermentation of organic matter in the soil. During the wet season, rice cannot hold the carbon in anaerobic conditions. The microbes in the soil convert the carbon into methane which is then released through the respiration of the rice plant or through diffusion of water. Current contributions of methane from agriculture is ~15% of anthropogenic greenhouse gases, as estimated by the IPCC. Further rise in sea level of 10-85 centimeters would then stimulate the release of more methane into the air by rice plants. Methane is twenty times more effective as a greenhouse gas than carbon dioxide.

## Pests and Diseases

Rice pests are any organisms or microbes with the potential to reduce the yield or value of the rice crop (or of rice seeds). Rice pests include weeds, pathogens, insects, rodents, and birds. A variety of factors can contribute to pest outbreaks, including the overuse of pesticides and high rates of nitrogen fertilizer application. Weather conditions also contribute to pest outbreaks. For example, rice gall midge and army worm outbreaks tend to follow high rainfall early in the wet season, while thrips outbreaks are associated with drought.

One of the challenges facing crop protection specialists is to develop rice pest management techniques which are sustainable. In other words, to manage crop pests in such a manner that future crop production is not threatened. Rice pests are managed by cultural techniques, pest-resistant rice varieties, and pesticides (which include insecticide). Increasingly, there is evidence that farmers' pesticide applications are often unnecessary . By reducing the populations of natural enemies of rice pests, misuse of insecticides can actually lead to pest outbreaks. Botanicals, so-called "natural pesticides", are used by some farmers in an attempt to control rice pests, but in general the practice is not common. Upland rice is grown without standing water in the field. Some upland rice farmers in Cambodia spread chopped leaves of the bitter bush (*Chromolaena odorata* (L.)) over the surface of fields after planting. The practice probably helps the soil retain moisture and thereby facilitates seed germination. Farmers also claim the leaves are a natural fertilizer and helps suppress weed and insect infestations.

Among rice cultivars there are differences in the responses to, and recovery from, pest damage. Therefore, particular cultivars are recommended for areas prone to certain pest problems. The genetically based ability of a rice variety to withstand pest attacks is called resistance. Three main types of plant resistance to pests are recognized as nonpreference, antibiosis, and tolerance. Nonpreference (or antixenosis) describes host plants which insects prefer to avoid; antibiosis is where insect survival is reduced after the ingestion of host tissue; and tolerance is the capacity of a plant to produce high yield or retain high quality despite insect infestation. Over time, the use of pest resistant rice varieties selects for pests that are able to overcome these mechanisms of resistance. When a rice variety is no longer able to resist pest infestations, resistance is said to have broken down. Rice varieties that can be widely grown for many years in the presence of pests, and retain their ability to withstand the pests are said to have durable resistance. Mutants of popular rice varieties are regularly screened by plant breeders to discover new sources of durable resistance.

Major rice pests include the brown planthopper (Preap et al. 2006), armyworms the green leafhopper, the rice gall midge, the rice bug, hispa, the rice leaffolder, stemborer, rats and the weed *Echinochloa crusgali*. Rice weevils are also known to be a threat to rice crops in the US, PR China and Taiwan.

Major rice diseases include Rice Ragged Stunt, Sheath Blight and Tungro. Rice blast, caused by the fungus *Magnaporthe grisea*, is the most significant disease affecting rice cultivation.

## Cultivars

While most breeding of rice is carried out for crop quality and productivity, there are varieties selected for other reasons. Cultivars exist that are adapted to deep flooding, and these are generally called 'floating rice'. The largest collection of rice cultivars is at the International Rice Research Institute (IRRI), with over 100,000 rice accessions held in the International Rice Genebank Rice cultivars are often classified by their grain shapes and texture. For example, Thai Jasmine rice is long-grain and relatively less sticky, as long-grain rice contains less amylopectin than short-grain cultivars. Chinese restaurants usually serve long-grain as plain unseasoned steamed rice. Japanese mochi rice and Chinese sticky rice are short-grain. Chinese people use sticky rice which is properly known as "glutinous rice" (note: glutinous refer to the glue-like characteristic of rice; does not refer to "gluten") to make zongzi. The Japanese table rice is a sticky, short-grain rice. Japanese sake rice is another kind as well.

Indian rice cultivars include long-grained and aromatic Basmati (grown in the North), long and medium-grained Patna rice and short-grained Sona Masoori (also spelled Sona Masuri). In South India the most prized cultivar is 'ponni' which is primarily grown in the delta regions of Kaveri River. Kaveri is also referred to as ponni in the South and the name reflects the geographic region where it is grown. In the Western Indian state of Maharashtra, a short grain variety called Ambemohar is very popular. this rice has a characteristic fragrance of Mango blossom.

Aromatic rices have definite aromas and flavours; the most noted cultivars are Thai fragrant rice, Basmati, Patna rice, and a hybrid cultivar from America sold under the trade name, Texmati. Both Basmati and Texmati have a mild popcorn-like aroma and flavour. In Indonesia there are also *red* and *black* cultivars. High-yield cultivars of rice suitable for cultivation in Africa and other dry ecosystems called the new rice for Africa (NERICA) cultivars have been developed. It is hoped that their cultivation will improve food security in West Africa.

Draft genomes for the two most common rice cultivars, *indica* and *japonica*, were published in April 2002. Rice was chosen as a model organism for the biology of grasses because of its relatively small genome (~430 megabase pairs). Rice was the first crop with a complete genome sequence. On December 16, 2002, the UN General Assembly declared the year 2004 the International Year of Rice. The declaration was sponsored by more than 40 countries.

### High-yielding Varieties

The High Yielding Varieties (HYV) are a group of crops created intentionally during the Green Revolution to increase global food production. Rice, like corn and wheat, was genetically manipulated to increase its yield. This project enabled labor markets in Asia to shift away from agriculture, and into industrial sectors. The first 'modern rice', IR8 was produced in 1966 at the International Rice Research Institute which is based in the Philippines at the University of the Philippines' Los Banos site. IR8 was created through a cross between an Indonesian variety named "Peta" and a Chinese variety named "Dee Geo Woo Gen."

With advances in molecular genetics, the mutant genes responsible for reduced height(rht), gibberellin insensitive (gai1) and slender rice (slr1) in Arabidopsis and rice were identified as cellular signaling components of gibberellic acid (a phytohormone involved in regulating stem growth via its effect on cell division) and subsequently cloned. Stem growth in the mutant background is significantly reduced leading to the dwarf phenotype. Photosynthetic investment in the stem is reduced dramatically as the shorter plants are inherently more stable mechanically. Assimilates become redirected to grain production, amplifying in particular the effect of chemical fertilizers on commercial yield. In the presence of nitrogen fertilizers, and intensive crop management, these varieties increase their yield 2 to 3 times.

## Potentials for the Future

As the UN Millennium Development project seeks to spread global economic development to Africa, the 'Green Revolution' is cited as the model for economic development. With the intent of replicating the successful Asian boom in agronomic productivity, groups like the Earth Institute are doing research on African agricultural systems, hoping to increase productivity. An important way this can happen is the production of 'New Rices for Africa' (NERICA). These rices, selected to tolerate the low input and harsh growing conditions of African agriculture are produced by the African Rice Center, and billed as technology from Africa, for Africa. The NERICA have appeared in *The New York Times* (October 10, 2007) and *International Herald Tribune* (October 9, 2007), trumpeted as miracle crops that will dramatically increase rice yield in Africa and enable an economic resurgence.

## Golden Rice

German and Swiss researchers have engineered rice to produce Beta-carotene, with the intent that it might someday be used to treat vitamin A deficiency. Additional efforts are being made to improve the quantity and quality of other nutrients in golden rice. The addition of the carotene turns the rice gold.

## Expression of Human Proteins

Ventria Bioscience has genetically modified rice to express lactoferrin, lysozyme, and human serum albumin which are proteins usually found in breast milk. These proteins have antiviral, antibacterial, and antifungal effects.

Rice containing these added proteins can be used as a component in oral rehydration solutions which are used to treat diarrheal diseases, thereby shortening their duration and reducing recurrence. Such supplements may also help reverse anemia.

### Others

The Chinese character for the *rice' kome*?) is composed by two eights *hachi*?) and ten ( which is 88, eighty-eight *hachi-jyu-hachi*?). In proverbial saying in Japan, the farmer spends eighty-eight times and efforts on rice from planting to crop and this is also teaching the sense of mottainai and gratitude for farmer and rice itself.

# 13

# Tobacco

## INTRODUCTION

Tobacco is an agricultural product, recognized as an addictive drug, processed from the fresh leaves of plants in the genus *Nicotiana*. Tobacco has long been in use as an entheogen in the Americas. However, upon the arrival of Europeans in North America, it quickly became popularized as a trade item and as a recreational drug. This popularization led to the development of the southern economy of the United States until it gave way to cotton. Following the American Civil War, a change in demand and a change in labor force allowed for the development of the cigarette. This new product quickly led to the growth of tobacco companies until the scientific controversy of the mid-1900s.

There are many species of tobacco, which are all encompassed by the plant genus *Nicotiana*. The word *nicotiana* (as well as *nicotine*) was named in honor of Jean Nicot, French ambassador to Portugal, who in 1559 sent it as a medicine to the court of Catherine de Medici. The effects of tobacco on human health are significant, and vary depending on the method by which it used and the amount consumed. Of the various methods of consumption the primary health risks pertain to diseases of the cardiovascular system by the vector of smoking, which over time allows high quantities of carcinogens to deposit in the mouth, throat, and lungs.

Because of the addictive properties of nicotine, tolerance and dependence develop. Absorption quantity, frequency, and speed of tobacco consumption are believed to be directly related to biological strength of nicotine dependence, addiction, and tolerance The usage of tobacco, is an activity that is practiced by some 1.1 billion people, and up to 1/3 of the adult population. The World Health Organisation reports it to be the leading preventable cause of death worldwide and estimates that it currently causes 5.4 million deaths per year. Rates of smoking have leveled off or declined in developed countries, however they continue to rise in developing countries.

Tobacco is cultivated similar to other agricultural products. Seeds are sown in cold frames or hotbeds to prevent attacks from insects, and then transplanted into the fields. Tobacco is an annual crop, which is usually harvested in a large single-piece farm equipment. After harvest, tobacco is stored to allow for curing, which allow for the slow oxidation and degradation of carotenoids. This allows for the agricultural product to take on properties that are usually attributed to the "smoothness" of the smoke. Following this, tobacco is packed into its various forms of consumption which include smoking, chewing, sniffing, and so on.

### Etymology

The Spanish word "*tabaco*" is thought to have its origin in Arawakan language, particularly, in the Taino language of the Caribbean. In Taino, it was said to refer either to a roll of tobacco leaves, or to the *tabago*, a kind of Y-shaped pipe for sniffing tobacco smoke (according to Oviedo; with the leaves themselves being referred to as *Cohiba*). However, similar words in Spanish and Italian were commonly used from 1410 to define medicinal herbs, originating from the Arabic *tabbaq*, a word reportedly dating to the 9th century, as the name of various herbs.

## EARLY DEVELOPMENTS

Tobacco had already long been used in the Americas when European settlers arrived and introduced the practice to Europe, where it became popular. At high doses, tobacco can become hallucinogenic; accordingly, Native Americans did not always use the drug recreationally. Instead, it was often consumed as an entheogen; among some tribes, this was done only by experienced shamans or medicine men Eastern North American tribes would carry large amounts of tobacco in pouches as a readily accepted trade item and would often smoke it in pipes, either in defined ceremonies that were considered sacred, or to seal a bargain, and they would smoke it at such occasions in all stages of life, even in childhood. It was believed that tobacco was a gift from the Creator and that the exhaled tobacco smoke was capable of carrying one's thoughts and prayers to heaven.

## Popularization

Following the arrival of the Europeans, tobacco became increasingly popular as a trade item. It fostered the economy for the southern United States until it was replaced by cotton. Following the American Civil War, a change in demand and a change in labor force allowed inventor James Bonsack to create a machine which automated cigarette production. This increase in production allowed tremendous growth in the tobacco industry until the scientific revelations of the mid-1900s.

## Temporary

Following the scientific revelations of the mid-1900s, tobacco became condemned as a health hazard, which eventually came to encompass as a cause for cancer, and other respiratory and circulatory diseases. This led to the Tobacco Master Settlement Agreement (MSA) which settled the lawsuit in exchange for a combination of yearly payments to the states and voluntary restrictions on advertising and marketing of tobacco products.

As the industry's downward tumble continued, in the 1990s Brown & Williamsons cross-bred a strain a tobacco to produce Y1. This strain of tobacco contained an unusually high amount of nicotine, nearly doubling its content from 3.2-3.5% to 6.5%. This prompted Food and Drug Administration (FDA) to use this strain as evidence that tobacco companies were intentionally manipulating the nicotine content of cigarettes.

In 2003, in response to tobaccos growth in developing countries, the World Health Organisation (WHO successfully rallied 168 countries to sign the Framework Convention on

Tobacco Control. The Convention is designed to push for effective legislation and its enforcement in all countries to reduce the harmful effects of tobacco. This led to the development of tobacco cessation products.

## NICOTIANA

There are many species of tobacco, which are encompassed by the genus of herbs *Nicotiana*. It is part of the nightshade family (Solanaceae) indigenous to North and South America, Australia, south west Africa and the South Pacific. Many plants contain nicotine, a powerful neurotoxin that is particularly harmful to insects. However, tobaccos contain a higher concentration of nicotine than most other plants. Unlike many other Solanaceae they do not contain tropane alkaloids, which are often poisonous to humans and other animals. Despite containing enough nicotine and other compounds such as germacrene and anabasine and other piperidine alkaloids (varying between species) to deter most herbivores a number of such animals have evolved the ability to feed on *Nicotiana* species without being harmed. Nonetheless, tobacco is unpalatable to many species and therefore some tobacco plants (chiefly Tree Tobacco, *N. glauca*) have become established as invasive weeds in some places.

### Primary Risks

The main health risks in tobacco pertain to diseases of the cardiovascular system, in particular smoking being a major risk factor for a myocardial infarction (heart attack), diseases of the respiratory tract such as Chronic Obstructive Pulmonary Disease (COPD) and emphysema, and cancer, particularly lung cancer and cancers of the larynx and mouth. It also increases the risk of developing pancreatic cancer by 75%. Prior to World War I, lung cancer was considered to be a rare disease, which most physicians would never see during their career. With the postwar rise in popularity of cigarette smoking came a virtual epidemic of lung cancer.

A number of studies have shown that tobacco use is a significant factor in spontaneous abortions among pregnant smokers, and that it contributes to a number of other threats to the health of the fetus. Second-hand smoke appears to present an equal danger to the fetus, as one study noted that "heavy *paternal* smoking increased the risk of early pregnancy loss." Second-hand smoke also presents a risk to infants, increasing their chances of Sudden Infant Death Syndrome (cot death).

## Epidemiology

The usage of tobacco is an activity that is practiced by some 1.1 billion people, and up to 1/3 of the adult population The World Health Organisation estimated in 2002 that in developed countries, 26% of male deaths and 9% of female deaths were attributable to smoking. Similarly, the United States Center for Disease Control and Prevention describes tobacco use as "the single most important preventable risk to human health in developed countries and an important cause of premature death worldwide. Rates of smoking have leveled off or declined in the developed world but continue to rise in developing countries. Smoking rates in the United States have dropped by half from 1965 to 2006, falling from 42% to 20.8% in adults. In the developing world, tobacco consumption is rising by 3.4% per year.

## ACTIVE SUBSTANCES

Long-term exposure to other compounds in the smoke, such as carbon monoxide, cyanide, and other compounds that damage lung and arterial tissue, are believed to be responsible for cardiovascular damage and for loss of elasticity in the alveoli, leading to emphysema and COPD. There are over 19 known carcinogens in cigarettes. The following are some of the most potent carcinogens:

- *Polynuclear aromatic hydrocarbons* are tar components produced by pyrolysis in smoldering organic matter and emitted into smoke.
- *Acrolein* is a pyrolysis product that is abundant is cigarette smoke. It gives smoke an acrid smell and an irritating, lachromatory effect and is a major contributor to its carcinogenity.
- *Nitrosamines* are a group of carcinogenic compounds found in cigarette smoke but not in uncured tobacco leaves.

In addition to chemical, nonradioactive carcinogens, tobacco and tobacco smoke contain small amounts of lead-210

(210Pb) and polonium-210 (210Po) both of which are radioactive carcinogens. The radioactive elements in tobacco are accumulated from the minerals in the soil, as with any plant, but are also captured on the sticky surface of the tobacco leaves in excess of what would be seen with plants not having this property. Research by NCAR radiochemist Ed Martell determined that radioactive compounds in cigarette smoke are deposited in "hot spots" where bronchial tubes branch. Since tar from cigarette smoke is resistant to dissolving in lung fluid, the radioactive compounds have a great deal of time to undergo radioactive decay before being cleared by natural processes.

### Benefits

Tobacco has sometimes been reported to have positive effects on certain medical conditions, presumably due to the biological effects of nicotine. Most notably, some studies have found that patients with Alzheimer's disease are more likely not to have smoked than the general population. However, the research in this area is limited and the results are conflicting, and some studies show that smoking increases the risk of Alzheimer's disease. A recent review of the available scientific literature concluded that the reduction of Alzheimer's disease may be confounded because smokers die before the age at which Alzheimer normally occurs.

A very large percentage of schizophrenics smoke tobacco as a form of self medication. The high rate of tobacco use by the mentally ill is a major factor in their decreased life expectancy, which is about 25 years shorter than the general population Following the observation that smoking improves condition of people with schizophrenia, in particular working memory deficit, nicotine patches had been proposed as a way to treat schizophrenia.

## CULTIVATION

Tobacco is cultivated similar to other agricultural products. Seeds were at first quickly scattered onto the soil. However, young plants came under increasing attack from flea beetles (*Epitrix cucumeris* or *Epitrix pubescens*), which caused destruction of half the tobacco crops in United States in 1876. By 1890 successful experiments were conducted that placed the plant in a frame covered by thin fabric.

Today, tobacco is sown in cold frames or hotbeds, as their germination is activated by light. After the plants have reached relative maturity, they are transplanted into the fields, in which a relatively large hole is created in the tilled earth with a tobacco peg.

In the United States, tobacco is often fertilized with the mineral apatite, which partially starves the plant of nitrogen to produces a more desired flavor. Apatite, however, contains radium, lead 210, and polonium 210—which are known radioactive carcinogens.

Tobacco is cultivated annual, and can be harvested in several ways. In the oldest method, the entire plant is harvested at once by cutting off the stalk at the ground with a sickle. In the nineteenth century, bright tobacco began to be harvested by pulling individual leaves off the stalk as they ripened. The leaves ripen from the ground upwards, so a field of tobacco may go through several so-called "pullings," more commonly known as topping (topping always refers to the removal of the tobacco flower before the leaves are systematically removed and, eventually, entirely harvested.

As the industrial revolution took hold, harvesting wagons used to transport leaves were equipped with man-powered stringers, an apparatus which used twine to attach leaves to a pole. In modern times large fields are harvested by a single piece of farm equipment, although topping the flower and in some cases the plucking of immature leaves is still done by hand.

## Curing

Curing and subsequent aging allows for the slow oxidation and degradation of carotenoids in tobacco leaf. This produces certain compounds in the tobacco leaves very similar and give a sweet hay, tea, rose oil, or fruity aromatic flavor that contribute to the "smoothness" of the smoke. Starch is converted to sugar which glycates protein and is oxidized into advanced glycation endproducts (AGEs), a caramelization process that also adds flavor. Inhalation of these AGEs in tobacco smoke contributes to atherosclerosis and cancer Levels of AGE's is dependent on the curing method used.

Tobacco can be cured through several methods which include but are not limited to:

- *Air cured* tobacco is hung in well-ventilated barns and allowed to dry over a period of four to eight weeks. Air-cured tobacco is low in sugar, which gives the tobacco smoke a light, sweet flavor, and high in nicotine. Cigar and burley tobaccos are air cured.
- *Fire cured* tobacco is hung in large barns where fires of hardwoods are kept on continuous or intermittent low smoulder and takes between three days and ten weeks, depending on the process and the tobacco. . Fire curing produces a tobacco low in sugar and high in nicotine. Pipe tobacco, chewing tobacco, and snuff are fire cured.
- *Flue cured* tobacco was originally strung onto tobacco sticks, which were hung from tier-poles in curing barns (Aus: kilns, also traditionally called Oasts). These barns have flues which run from externally fed fire boxes, heat-curing the tobacco without exposing it to smoke, slowly raising the temperature over the course of the curing. The process will generally take about a week. This method produces cigarette tobacco that is high in sugar and has medium to high levels of nicotine.
- *Sun-cured* tobacco dries uncovered in the sun. This method is used in Turkey, Greece and other Mediterranean countries to produce oriental tobacco. Sun-cured tobacco is low in sugar and nicotine and is used in cigarettes.

## Products

Tobacco can be processed into a number of products which include but are not limited to:

- *Chewing tobacco*, Chewing is one of the oldest ways of consuming tobacco leaves. It is consumed orally, but is rarely consumed though actual chewing. Small amounts are placed at the bottom lip, between the gum and the teeth, where it is gently compacted, thus it can oftentimes be encompass Dipping tobacco. This stimulates the salve glands, which led to the development of the spittoon.

- *Creamy snuff*, is a tobacco paste, consisting of tobacco, clove oil, glycerin, spearmint, menthol, and camphor, and sold in a toothpaste tube. It is marketed mainly to women in India, and is known by the brand names Ipco (made by Asha Industries), Denobac, Tona, Ganesh. It is locally known as "mishri" in some parts of Maharashtra.
- *Dipping tobacco*, is a form of smokeless tobacco. Dip is occasionally referred to as "chew", and because of this, it is commonly confused with chewing tobacco, which encompasses a wider range of products. A small clump of dip is 'pinched' out of the tin and placed between the lower or upper lip and gums.
- *Gutka*, (also spelled gutkha, guttkha, guthka) is a preparation of crushed betel nut, tobacco, and sweet or savory flavorings. It is manufactured in India and exported to a few other countries. A mild stimulant, it is sold across India in small, individual-size packets.
- *Snuff* is a generic term for fine-ground smokeless tobacco products. Originally the term referred only to dry snuff, a fine tan dust popular mainly in the eighteenth century. Snuff powder originated in the UK town of Great Harwood and was famously ground in the town's monument prior to local distribution and transport further up north to Scotland. There are two major varieties which include European (dry) and American (moist); although American snuff is often referred to as dipping tobacco.
- *Snus*, is a moist powder tobacco product that is consumed by placing it under the upper lip for extended periods of time. It is a form of snuff that is used in a manner similar to American dipping tobacco, but typically does not result in the need for spitting.
- *Topical tobacco* paste is sometimes recommended as a treatment for wasp, hornet, fire ant, scorpion, and bee stings. An amount equivalent to the contents of a cigarette is

mashed in a cup with about a 0.5 to 1 teaspoon of water to make a paste that is then applied to the affected area.

- *Tobacco water* is a traditional organic insecticide used in domestic gardening. Tobacco dust can be used similarly. It is produced by boiling strong tobacco in water, or by steeping the tobacco in water for a longer period. When cooled the mixture can be applied as a spray, or 'painted' on to the leaves of garden plants, where it will prove deadly to insects.

## Cultural

Due to its long existence, tobacco has fostered many cultural items including: the usage of peace pipes, advertisements, movies. From its discovery tobacco has been highly regarded used in cultural ceremonies, for recreational purposes, and so forth. At the arrivals of the Europeans, tobacco was came to be regarded with wealth and knowledge. Smoking in public has for a long time been something reserved for men and when done by women has been associated with promiscuity. In Japan during the Edo period, prostitutes and their clients would often approach one another under the guise of offering a smoke and the same was true for 19th century Europe.

Following the Civil War, the usage of tobacco, primarily in cigarettes became associated with masculinity, and power and is an iconic image associated with the stereotypical capitalist.From the mid-1900s and to today Tobacco is often rejected. This has spawned quitting-associations, negative-campaigns, and so forth. Bhutan is the only country in the world where tobacco sales are illegal.

## Plantation Economy

A plantation economy is an economy which is based on agricultural mass production, usually of a few staple products grown on large farms called plantations. Plantation economies rely on the export of cash crops as a source of income. Prominent plantation crops have included cotton, rubber, sugar cane, tobacco, figs, rice, kapok, sisal and indigo.

Regions with plantation economies have usually been in the southern United States, South America, the Caribbean, and Africa. Fordlândia is a 20th century example of a plantation economy. Plantation economies are also historically associated with slavery, particularly in the United States. Plantation economies usually benefit the large countries to which they are exporting, which usually manufacture the raw materials grown on the plantations into goods which are then traded back to the plantation economy. Throughout most of history, the countries receiving the crops have usually been in Western Europe.

## United States

Tobacco production was labor intensive and required thousands of slaves to produce millions of pounds that were exported. The period covered by this article ranges from 1700 to the end of the American Civil War (1865). The wealth and Influence of the so-called "tuckahoe" Virginia planters depended on one crop, and that crop was tobacco. The production of tobacco spread down the James, York, Rappahannock, and the Potomac rivers.

## Tobacco and Virginia Economy

Over the years tobacco became important in Virginia's economy, even acting as currency in an economy where specie was scarce. An independent currency allowed the colonies to gain power and slowly break away from the British economically and culturally. In the year 1758 Virginia exported 70,000 hogsheads of tobacco. The production of tobacco in colonial times required much toil. The plants had to be grown from seeds in a cold frame, set out, weeded, tasseled, harvested, and cured. All of this work was done by man and beast. Each acre produced about 5000 plants that required hand care over and over again. But, with slave labour, profits exceeded any other plant that could be grown.

Many of the wealthy and influential men in Colonial Virginia were Planters; tobacco plantation owners. A number of America's first presidents owned slaves. They owned numerous plantations, each with large numbers of slaves.

## Slave Statistics

In the year 1860, one out of every four families in Virginia owned slaves. The figures cited here are from the 1860 census. There were over 100 plantation owners that owned over 100 slaves.

## Slave being Inspected

Number of slaves in the Lower South: 2,312,352 (47% of total population).

Number of slaves in the Upper South: 1,208,758 (29% of total population).

Number of slaves in the Border States: 432,586 (13% of total population).

Fewer than one-third of all Southern families owned slaves at the peak of slavery prior to the Civil War. In Mississippi and South Carolina it approached one half. The total number of slave owners was 385,000 (including, in Louisiana, some free Negroes), amounting to approximately 3.8% of the Southern and Border states population.

On a typical plantation of more than 20 slaves, the capital value of the slaves was greater than the capital value of the land and implements.

Sugar has a long history as a plantation crop. Growing had to follow a precise, scientific system in order to profit from the production. Sugar plantations everywhere were disproportionate consumers of labour—often enslaved—owing to the high mortality of the plantation laborers.

The slaves working the sugar plantation were caught in an unceasing rhythm of arduous labor year after year. Sugarcane is harvested about 18 months after planting and the plantations usually divided their land for efficiency. One plot was lying fallow, one plot was growing cane, and the final plot was being harvested. During the May-December rainy season, slaves planted, fertilized with animal dung, and weeded. From January to June, they harvested the cane by chopping the plants off close to the ground, stripping the leaves, then cutting them into shorter strips to be bundled off to be sent to the mill.

In the mill, the cane was crushed using a three roller mill. The juice from the crushing of the cane was then boiled or clarified until it crystallized into sugar. Some plantations also went a step further and distilled the molasses (the liquid left after the sugar is boiled or clarified) to make rum. The sugar was then shipped back to Europe, and for the slave laborer the routine started all over again. With the 19th century abolition of slavery, plantations continued to grow cane, but sugar beets not grown on plantations increased their market share.

# 14

# Economy from Bee Plants

## INTRODUCTION

When most people think of bees, the first bee that comes to mind is the honey bee. But this bee is only one of about 25,000 species known worldwide. In the U.S. we have almost 4,000 and in California slightly more than 1,500 species have been recorded. One reason for high species diversity in our state is the diverse flowering plants that can be found in a wide diversity of habitat types. Focusing down on the San Francisco Bay Area, we have already identified 81 species of bees from the following five families:

- Apidae;
- Andrenidae;
- Colletidae;
- Halictidae; and
- Megachilidae.

It takes some study and practice, usually with a trained professional, before one can begin to identify bees to the family level, and then to the more difficult genus and species level. In our scientific work we strive to identify all bees to the species level, but we hope that the following general descriptions of the five families as well as the most common genera and species found in Bay Area gardens will help to open your eyes to the many interesting bees that are pollinating your flowers. We have also included references for more detailed descriptions of bee types.

*Apids* constitute a highly diverse family of bees, which include social, hive-building groups like the honey bee as well as solitary groups which make individual nests in the ground or in tree holes:

- The most common social member of the Apids is the honey bee, *Apis mellifera,* which is an introduced (exotic) species from Europe. Honey bees measure about 3/4 of an inch in length. They are easily recognized by their slow flight and flower visitation. They come in a range of colors from blond to black and usually have brown bands on their abdomens. This bee carries pollen neatly in 'pollen baskets' or hairs specially developed to collect pollen on their hind legs.
- Bumble bees, or *Bombus* species, are another common social group of apid bees in the Bay Area. They are often larger, rounder, and have more hair than honey bees. They are usually black with 1 or 2 yellow stripes, and are called 'bumble bees' because of their bumbling flight pattern. Like the honey bee, this bee group carries pollen, sometimes in large yellow or purple blobs, in pollen baskets on its hind legs.
- *Anthophorids* or digger bees are medium-sized compared to honey bees. They are fast flyers and visit flowers rapidly and efficiently compared to honey bees. They are often quite hairy, and females appear roundish when their hind legs are packed with pollen. the most common genus of anthophorid is *Melissodes,* which contains two species, both of which are a little smaller than honey bees. Males (with long bodies and antennae) and females (roundish bodies and short antennae) are commonly found during the summer on flowers of *Cosmos* spp., *Helianthus annuus, Scabiosa* spp., *Coreopsis* spp., and *Bidens ferulifolia* (photos). At night male *Melissodes* can often be found sleeping together in flowers of *Cosmos bipinatus.*

Another prominent, but less common anthophorid, is *Anthophora urbana* . It is a fast flying, compact bee about the size of *Melissodes* females. It has a grayish thorax and a distinct black and white banded abdomen and visits a wide variety of ornamental flowers in spring and summer.

- Carpenter bees, or *Xylocopids,* look a lot like bumble bees, especially with regard to size and their rounded body shape. They differ from bumbles in that they are mostly solid, shiny black, have little hair, and the top of the abdomen is flattened.
- *Ceratina,* or small carpenter bees measure about 1/4 to 1/2 of an inch in length. Their bodies are long and slender and are metallic blue, purple or black in colour. They have no obvious hair. Observing these skittish little bees requires some skill, however, once the general form and color of the abdomen are recognized, there is no mistaking them.

Andrenids are solitary spring bees, small to medium sized compared to honey bees, and also carry neatly-packed pollen on their hind legs. Most of the Bay Area species are shiny blue-black with torpedo-shaped abdomens, but some have beige or brown hair covering their thoraxes. These bees are not as common as other spring bees, however, we have observed them with some consistency on *Phacelia tanacetifolia, Gilia capitata,* and some chard and kale plants.

*Colletids* are regarded by some taxonomic specialists as a group of primitive bees. They are mostly solitary in habit. There are two subfamilies and members of both are found in the Bay Area. Due to their small size, *Colletids* are very difficult to photograph.

- Species of the *Colletinae* subfamily are mostly robustly hairy bees, up to 3/4 inch in length, which resemble anthophorid bees in general appearance.
- Bees of the *Hylaeinae* subfamily are very small and wasp-like, about 1/2 an inch in length. Most species are black with conspicuous yellow markings on their faces. In contrast to other bee groups, females of this subfamily carry pollen in the crop, which is the specialized anterior portion of the digestive tract.

*Halictids* are common spring and summer bees, small to medium in size, with long bodies. They are very diverse in lifestyles, including solitary, semi-social, and primitively social species. As a group, the halictids are commonly observed visiting a wide variety of ornamental flowers. Upon close examination at pollen flowers such as California poppy (*Eschscholzia californica*), females can be observed gathering pollen on their legs, especially the hind legs. During this process, they will also commonly have pollen messily spattered on the sides of the thorax and abdomen, as opposed to the neat pollen-carrying baskets of the honey and bumble bees.

- The group *Halictus*, like most species in the Bay Area, have elongated abdomens that are black with light coloured bands.
- The Females of the colorful species *Agapostemon texanus* are metallic green, and the males have metallic green thoraxes and yellow and black striped abdomens. Females of this group carry pollen on their legs, but often the pollen is also loosely affixed to forelegs and the sides of the abdomen.

*Megachilids* are known as leaf cutting and mason bees. They are a large, diverse, and common group of bees found in the Bay Area. They range in size from small (1/2 an inch) to honey bee size, and most are strictly solitary. The most common Bay Area species are dark gray with distinct light-colored banding on the abdomen. In sharp contrast to the other families, megachilid females gather pollen on the underside of the abdomen by means of a series of stout hairs that protrude from several body segments. While visiting flowers, some female megachilids arch their backs, elevating their abdomens and showing their densely-packed and often colorful pollen loads. Although megachilids are found during the entire growing season, certain groups can only be found during specific seasons.

- One group, *Osmia*, can be observed only during the spring and early summer period. *Osmia* are shiny metallic blue-black, bluish, or green depending on the species. They are a delight to observe rapidly visiting ornamental flowers such as *Salvia mellifera* and *Clarkia unguiculata*.

- The largest Bay Area megachilid is *Megachile perihirta* (pictured above-right). It is about the same size as a honey bee. It can be observed at close range from spring to early fall visiting a wide variety of plants, and especially *Coreopsis* species, *Cosmos bipinnatus, C. sulphureus,* and *Helianthus annuus.*

During the late summer-early fall another large megachilid, *Chalicodoma* species, can be occasionally observed on flowers of *Bidens ferulifolia, Cosmos bipinnatus,* and *Helianthus annuus.* This bee is easily recognized by the elegant color pattern of gold and black bands on its abdomen. (photo on cosmos).

"Small and medium bees" Some plants such as *Calendula* species, *Geranium incanum* "Sugar Plum", or the buckwheats characteristically attract small or medium sized bees (photos). We are not always able to identify those bees on flowers when we make field observation. For convenience we refer to these plants as having small or medium bees until we have enough information for at least a family name for these groups of bees. In the case of the small bees, they often turn out to be members of the family *Halictidae* or rarely

## MEDIUM BEES USUALLY ARE MEGACHILIDS

### General Info on Our Plant Recommendations

We refer to bee plants as those that have measurable attraction to urban bees. These are plants that are regularly visited by bees for their pollen and nectar resources.

The bee plants listed in this section are the ones that we have evaluated in the San Francisco East Bay Area for their attractiveness. We have listed mostly the smaller plants that can be grown rather quickly in gardens. Trees and large shrubs attractive to bees are not covered here – they can be found in the General List of known bee plants.

In the following Most Attractive Plant List we provide only the essential characteristics of the bee plants, which should be useful for anyone planning a bee or pollinator garden. We advise readers to consult the *Sunset Western Garden Book* (2001) for more complete information on the plants, vegetative characteristics, their history, variations, and preferred soils and other preferred growing conditions.

We expect this list of recommended plants to grow as we learn more about the floral preferences and behaviours of urban bees. We also continue to learn about new plants and new varieties of our listed plants that are attractive to bees. In this regard, several gardeners and nursery persons call new discoveries to our attention, which we follow up with evaluations whenever possible.

Regarding visitors, you will notice that with each plant we have provided major categories of associated bees and other flower visitors. Descriptions of the bee groups can be found below. At times we refer to flower-visiting bees as sb (small bee), mb (medium bee), or lb (large bee). This occurs when flower visits are too rapid or we fail to net the bees for more accurate taxonomic determination. As time goes on and we make more observations, these vague bee-size categories will be changed to actual taxonomic groups (section of web site).

Many of the bee plants also attract a wide variety of other flower visitors. When this occurs with regularity, we list these visitors as major groups such as: beneficial flies (usually hover flies); beneficial wasps (many are carnivorous species looking for prey insects feeding on plants, but some visit for nectar); and butterflies seeking floral nectar. A few of the bee plants also attract hummingbirds, such as *Salvia uliginosa* (pictured at right). Needless to say, with careful planning, gardens can become important micro-habitats for a wide variety of beneficial pollinators. In fact, by concentrating known pollinator plants in a small area, one can actually experience more insect life than in a wildland area where flower visitors are more widely spread over a landscape.

# 15

# Spice

## INTRODUCTION

A spice is a dried seed, fruit, root, bark or vegetative substance used in nutritionally insignificant quantities as a food additive for the purpose of flavoring, and sometimes as a preservative by killing or preventing the growth of harmful bacteria. Many of these substances are also used for other purposes, such as medicine, religious rituals, cosmetics, perfumery or eating as vegetables. For example, turmeric is also used as a preservative; licorice as a medicine; garlic as a vegetable. In some cases they are referred to by different terms. In the kitchen, spices are distinguished from herbs, which are leafy, green plant parts used for flavoring purposes. Herbs, such as basil or oregano, may be used fresh, and are commonly chopped into smaller pieces. Spices, however, are dried and often ground or grated into a powder. Small seeds, such as fennel and mustard seeds, are used both whole and in powder form.

## CLASSIFICATION AND TYPES

Spices can be grouped as:

- Dried fruits or seeds, such as fennel, mustard, and black pepper.
- Arils, such as mace.
- Barks, such as cinnamon and cassia.

- Dried buds, such as cloves.
- Stamens, such as saffron.
- Roots and rhizomes, such as turmeric, ginger and galingale.
- Resins, such as a foetida.

Herbs, such as bay, basil, and thyme are not, strictly speaking, spices, although they have similar uses in flavouring food. The same can be said of vegetables such as onions and garlic.

## Early History

The spice trade developed throughout the Middle East in around 2000 BC with cinnamon, Indonesian cinnamon and pepper. A archaeological discovery suggests that the clove, indigenous to the Indonesian island of Ternate in the Maluku Islands, could have been introduced to the Middle East very early on. Digs found a clove burnt onto the floor of a burned down kitchen in the Mesopotamian site of Terqa, in what is now modern-day Syria, dated to 1700 BC.

In the story of Genesis, Joseph was sold into slavery by his brothers to spice merchants. In the biblical poem Song of Solomon, the male speaker compares his beloved to many forms of spices. Generally, Egyptian, Chinese, Indian and Mesopotamian sources do not refer to known spices. In South Asia, nutmeg, which originates from the Banda Islands in the Moluccas, has a Sanskrit name. Sanskrit is the language of the sacred Hindu texts, this shows how old the usage of this spice is in this region. Historians estimate that nutmeg was introduced to Europe in the 6th century BC.

The ancient Indian epic of Ramayana mentions cloves. In any case, it is known that the Romans had cloves in the 1st century AD because Pliny the Elder spoke of them in his writings. Indonesian merchants went around China, India, the Middle East and the east coast of Africa. Arab merchants controlled the routes through the Middle East and India until Roman times with the discovery of new sea routes. This made the city of Alexandria in Egypt the main trading centre for spices because of its port.

## Middle Ages

Spices were among the most luxurious products available in Europe in the Middle Ages, the most common being black pepper, cinnamon (and the cheaper alternative cassia), cumin, nutmeg, ginger and cloves. They were all imported from plantations in Asia and Africa, which made them extremely expensive. From the 8th until the 15th century, the Republic of Venice had the monopoly on spice trade with the Middle East, and along it with the neighboring Italian city-states. The trade made the region phenomenally rich. It has been estimated that around 1000 tons of pepper and 1000 tons of the other common spices were imported into Western Europe each year during the Late Middle Ages. The value of these goods was the equivalent of a yearly supply of grain for 1.5 million people. While pepper was the most common spice, the most exclusive was saffron, used as much for its vivid yellow-red color as for its flavor. Spices that have now fallen into some obscurity include grains of paradise, a relative of cardamom which almost entirely replaced pepper in late medieval north French cooking, long pepper, mace, spikenard, galangal and cubeb. A popular modern-day misconception is that medieval cooks used liberal amounts of spices, particularly black pepper, merely to disguise the taste of spoiled meat. However, a medieval feast was as much a culinary event as it was a display of the host's vast resources and generosity, and as most nobles had a wide selection of fresh or preserved meats, fish or seafood to choose from, the use of ruinously expensive spices on cheap, rotting meat would have made little sense. In the Caribbean, the island of Grenada is well known for growing and exporting a number of spices including the nutmeg which was introduced to Grenada by the settlers.

## Early Modern Period

The control of trade routes and the spice-producing regions were the main reasons that Portuguese navigator Vasco da Gama sailed to India in 1499. Spain and Portugal were not happy to pay the high price that Venice demanded for spices. At around the same time, Christopher Columbus returned from the New World, he described to investors the many new, and then unknown, spices available there.

It was Afonso de Albuquerque (1453-1515) who allowed the Portuguese to take control of the sea routes to India. In 1506, he took the island of Socotra in the mouth of the Red Sea and, in 1507, Ormuz in the Persian Gulf. Since becoming the viceroy of the Indies, he took Goa in India in 1510, and Malacca on the Malay peninsula in 1511. The Portuguese could now trade directly with Siam, China and the Moluccas. The Silk Road complemented the Portuguese sea routes, and brought the treasures of the Orient to Europe via Lisbon, including many spices.

## COMMON SPICE MIXTURES

- Berbere (Ethiopia and Eritrea).
- Colombo (paprika, cumin, coriander, nutmeg, ginger, black pepper, star anise, cardamom, cloves, mustard grains, saffron).
- Curry powder (Indian-style, used in the West and Japan).
- Five bays.
- Five-spice powder (China).
- Herbes de Provence (Southern France).
- Jerk spice (Jamaica).
- Khmeli suneli (Georgia).
- Masalas, including garam masala (India).
- Old Bay Seasoning (United States).
- Panch phoron.
- Poultry Seasoning (United States).
- Pumpkin pie spice (United States).
- Quatre épices (France).
- Ras el hanout (Middle East/North Africa).
- Shichimi togarashi (Japan).
- Za'atar (Middle East).

## List of Culinary Herbs and Spices

This is a list of culinary herbs and spices. Specifically these are food or drink additives of mostly botanical origin used in nutritionally insignificant quantities for flavoring or coloring. As such, this list does not contain salt, which is a mineral, herbs or spices that are purely medicinal such as valerian, fictional plants such as aglaophotis, or recreational drugs such as marijuana. For medicinal herbs see list of medicinal herbs.

### A

Ajwain (*Trachyspermum ammi*)

Alkanet (*Anchusa arvensis*), mostly used as a natural colouring

Allspice (*Pimenta dioica*)

Almond

Alpinia galanga

Amchur - mango powder (Mangifera)

Angelica (*Angelica archangelica*)

Anise (*Pimpinella anisum*)

Aniseed myrtle (*Syzygium anisatum*)

Annatto (*Bixa orellana L.*)

Apple mint (*Mentha suaveolens*)

Mugwort (*Artemisia vulgaris*)

Asafoetida (*Ferula assafoetida*)

### B

Basil (*Ocimum basilicum*)

Bay Leaf

### C

Calendula

Calumba (*Jateorhiza calumba*)

Cananga Also known as 'Kenanga' in Malaysia.

Chamomile

Candle nut

Caper (*Capparis spinosa*)

Caraway

Cardamom

Carob Pod

Cassia

Casuarina

Catnip

Cat's Claw

Catsear

Cayenne pepper

Celastrus paniculatus (peng)

Celery salt

Celery seed

Centaury

Chervil (*Anthriscus cerefolium*)

Chickweed

Chicory (*Cichorium*)

Chilli pepper

Chipotle

Chives (*Allium schoenoprasum*)

Cicely (*Myrrhis odorata*)

Cilantro (see Coriander) (*Coriandrum sativum*)

Cinchona (Cinchona)

Cinnamon (and Cassia)

Cinnamon Myrtle (*Backhousia myrtifolia*)

Clary (*Salvia sclarea*)

Cleavers

Clover

Cloves

Coffee

Comfrey

Common Rue

Condurango

Coptis

Coriander

Costmary (*Tanacetum balsamita*)

Couchgrass

Cow Parsley (*Anthriscus sylvestris*)

Cowslip (*Primula veris*)

Cramp Bark (*Viburnum opulus*)

Cress

Cuban Oregano (*Plectranthus amboinicus*)

Cubeb pepper (*Piper cubeba*)

Cudweed

Culantro (*Eryngium foetidum*)

Cumin

Curry leaf (*Murraya koenigii*)

**D**

Damiana (*Turnera aphrodisiaca, T. diffusa*)

Dandelion (*Taraxacum officinale*)

Demulcent

Devil's claw (*Harpagophytum procumbens*)

Dill seed

Dill (Anethum graveolens)

Dorrigo Pepper (Tasmannia stipitata)

**E**

Echinacea -

Echinopanax elatum

Edelweiss

Elderberry

Elderflower

Elecampane

Eleutherococcus senticosus

Emmenagogue

Epazote (Chenopodium ambrosioides)

Ephedra -

Eryngium foetidum

Eucalyptus

Eyebright

**F**

Fennel (Foeniculum vulgare)

Fenugreek

Feverfew

Figwort

Filé powder

Fingerroot (Boesenbergia rotunda)

Fo ti ticng

French sorrel (Rumex scutatus)

Fumitory

**G**

Galangal

Galingale

Garden cress

Garlic chives

Garlic

Ginger (*Zingiber officinale*)

Ginkgo biloba

Ginseng

Ginseng, Siberian (*Eleutherococcus senticosus*)

Goat's rue (*Galega officinalis*)

Goada masala

Gotu Kola

Grains of paradise (*Aframomum melegueta*)

Grains of Selim (*Xylopia aethiopica*)

Grape seed extract

Green tea

Ground ivy (*Glechoma hederacea*)

Guaco

Gypsywort (*Lycopus europaeus*)

**H**

Hawthorn (Crataegus)

Herbes de Provence

Hibiscus

Holly (Ilex)

Holy Thistle

Hops, the female flower cones of hop (Humulus lupulus)

Horehound

Horseradish

Horsetail (*Equisetum telmateia*)

Hyssop (*Hyssopus officinalis*)

**I**

Imli (Tamarind)

**J**

Jalap

Jasmine

Javitri (mace)

Jiaogulan (*Gynostemma pentaphyllum*)

Joe Pye weed (Gravelroot)

John the Conqueror

Juniper

**K**

Kaffir Lime Leaves (*Citrus hystrix, C. papedia*)

Kaala masala

Knotweed (Polygonum)

Kokam

**L**

Labrador tea

Lady's Bedstraw (*Galium verum*)

Lady's Mantle (*Alchemilla*)

Land cress

Lavender (*Lavandula sp.*)

Ledum

Lemon Balm (*Melissa officinalis*)

Lemon basil

Lemongrass (*Cymbopogon citratus, C. flexuosus* and other species)

Lemon Ironbark (*Eucalyptus staigeriana*)

Lemon mint

Lemon Myrtle (*Backhousia citriodora*)

Lemon Thyme

Lemon verbena (*Lippia citriodora*)

Licorice - adaptogen

Lime Flower

*Limnophila aromatica*

Lingzhi

Linseed

Liquorice

Liverwort

Long pepper

Lovage (*Levisticum officinale*)

Luohanguo

**M**

Mace

Mahlab

Malabathrum

Manchurian Thorn Tree (*Aralia manchurica*)

Marjoram (*Origanum majorana*)

Marrubium vulgare (*white horehound*)

Marsh Labrador Tea

Marshmallow

Mastic

Meadowsweet (*Filipendula vulgaris*)

Mei Yen

Melegueta pepper ( *Aframomum melegueta*)

Mexican pepperleaf (*Piper auritum*)

Mint (Mentha spp.)

Milk thistle (Silybum) Bergamot (*Monarda didyma*)

Motherwort

Mountain Skullcap

Mullein (*Verbascum thapsus*)

*Murraya koenigii* (Curry Tree)

Mustard

Mustard seed

**N**

Nashia inaguensis

Neem

Nepeta (catmint)

Nettle (Urtica)

Nigella sativa (Kalonji, Black caraway)

Noni

Nutmeg (and Mace)

**O**

Oenothera (*Oenothera biennis* and other sp.)

Oilseed

Olida (*Eucalyptus olida*)

Oregano (*Origanum vulgare, O. heracleoticum,* and other species)

Orris root

Osmorhiza

Olive Leaf (used in tea and as herbal supplement)

**P**

Pandan leaf

Paprika

Paracress

Parsley (*Petroselinum crispum*)

Passion Flower (Passiflora)

Patchouli

Pennyroyal (*Mentha pulegoides*)

Pepper (black, white, and green)

Peng (*Celastrus paniculatus*)

Peppermint (*Mentha piperata*)

Peppermint Gum (*Eucalyptus dives*)

Perilla

Plantain

Pomegranate

Ponch phoran

Poppy seed

Primrose (*Primula vulgaris*)

Psyllium

Purslane

**Q**

Quassia

**R**

Ramsons (wood garlic, Allium ursinum)

Ras el-hanout

Raspberry (leaves)

Reishi

Restharrow (Ononis)

Rhodiola rosea

Riberry (*Syzygium luehmannii*)

Rocket/Arugula

Roman chamomile

Rooibos

Rosehips

Rosemary (*Rosmarinus officinalis*)

Rowan berries (*Sorbus aucuparia*)

Rue (Ruta)

**S**

Safflower

Saffron

Sage (*Salvia officinalis*)

Saigon Cinnamon

St John's Wort (Hypericum)

Salad Burnet (*Sanguisorba minor*)

Salvia (clary and sage)

Sassafras

Savory (Summer savory, Satureja hortensis, and winter savory, *S. montana*)

Schisandra chinensis

Scutellaria costaricana (skullcap)

Senna (herb)

*Senna obtusifolia*

Sesame seed

Sheep's sorrel (*Rumex acetosella*)

Shepherd's Purse

Sialagogue

Siberian Chaga

Siberian ginseng (*Eleutherococcus senticosus*)

Sichuan pepper (*Xanthoxylum piperitum*)

*Siraitia grosvenorii* (luohanguo)

Skullcap (Scutellaria)

Sloe berries (*Prunus spinosus*)

Smudge Stick

Sonchus (sow-thistle)

Sorrel (*Rumex acetosa, R. acetosella, R. scutatus*)

Southernwood

Spearmint

Speedwell

Squill (Scilla)

Star anise

Stevia

Strawberry leaves (Fragaria)

Suma (*Pfaffia paniculata*)

Sumac

Summer savory (*Satureja hortensis*)

Sutherlandia frutescens

Sweet grass

Sweet cicely (*Myrrhis odorata*)

Sweet woodruff

**T**

Tacamahac

Tamarind

Tansy

Tarragon (*Artemisia dracunculus*)

Tea (*Camellia sinensis*)

*Teucrium polium*

Thai basil

Thistle

Thyme

Toor Dall

Tormentil (*Potentilla erecta*)

*Tribulus terrestris*

Tulsi (*Ocimum tenuiflorum*)

Turmeric (*Curcuma longa*)

Twinleaf onion

**U**

Uva ursi (bearberry)

**V**

Vanilla (*Vanilla planifolia*)

Vasaka

Vervain

Vetiver

Vietnamese Coriander (*Persicaria odorata*)

**W**

Wasabi (*Wasabia japonica*)

Watercress (*Rorippa nasturtium-aquatica*)

Wattleseed

Wild ginger

Wild Lettuce

Wild thyme

Winter savory (*Satureja montana*)

Witch Hazel (*Hamamelis*)

Wolfberry

Wood Avens

Wood Betony

Woodruff

Wormwood

**Y**

Yarrow, (*Alchemilla millefolium*)

Yerba Buena

Yohimbe

Yomogi

**Z**

Zedoary root

Spice mixtures

Berbere

Curry powder

Five-spice powder or Chinese five-spice

Garam masala

Herbes de Provence

Khmeli Suneli

# 16

# Plants and the Environment

## INTRODUCTION

As plants interact with the environment directly by exchanging water and energy, they are very sensitive to storms, droughts and floods. These weather events can severely damage crop yield. In this unit we look at how plants are affected by changes in temperature, humidity and rainfall. Probably the most important role of plants in the environment is the production of oxygen ($O_2$) and the absorption of carbon dioxide ($CO_2$) from the atmosphere during the process of photosynthesis. Photosynthesis is the basic process for plant life.

Plants also affect and change their surroundings to make them more suitable for growth. Environmental conditions, such as light intensity, temperature, water availability and wind strength, affect plant growth. Plants also modify the environment around them, they release water which cools the air, breakdown the soil to make it suitable for their roots and for other plants and animals and decrease the speed of the wind. As you know, plants need water to live and grow. High temperatures reduce the availability of water and decrease crop yields. Young plants are especially vulnerable to extreme weather conditions.

Let´s now look at some of the mechanisms plants use to cope with extreme weather events such as high temperatures, floods or droughts. High temperatures affect crops directly by increasing the rate at which they loose water (their evaporation rate), in the same way that high temperatures make us sweat. Plants have very small pores, called stomata, spread over their leaves. These stomata help the plants control how much water they contain. In the picture you can see that the stomata are made up of two guard cells which open or close the pore depending on how much water the plant needs. During dry periods, the guard cells are closed so that the plants do not loose too much water. Under normal weather conditions, the stomata are open.

Each plant type has different structural characteristics and so all plants don't grow well at the same temperature. When the optimal temperature value for a particular plant is exceeded, the plant tends not to grow as well leading to a drop in yield. Most plants are very sensitive to high temperatures, although the extent varies depending on the age of the plant and its ability to withstand poor situations. High temperatures also increase the amount of water lost from the soil by evaporation. This also affects plant growth as the soil is their main source of water.

## PRECIPITATION

Precipitation (rainfall) is the primary source of soil moisture, and rainfall amount is probably the most important factor determining the crop yield. A change in climate can cause an increase or a decrease in the amount of precipitation which falls.

### Plant Roots

Roots are one of the main ways plants get water from the environment. In many parts of the world, plants have much longer roots than trunks or branches. Sometimes a bush that is only around 30 cm high can have roots that go as deep as two meters into the soil. This happens in places where there is not much rain during the year, like the deserts or the very arid dry regions of the world.

Plant roots spread throughout the soil so that water can be taken up in different places at the same time. The main roots have fine root hairs on them to allow water to be absorbed easily.

Dry periods can badly affect plant growth but the amount of damage depends on the ability of the plant to expand its root system and how much water the soil can hold onto. High humidity, frost and hail may also damage certain crops.

High temperatures are normally accompanied by dry periods and both the warmth and the lack of water can have negative effects on plant growth. Roots are unable to find water in the soil and stomata have to close to prevent water loss from the plant. This causes the temperature of the plant to rise and this can damage the plant. When plants struggle to grow because of high temperatures or lack of water, they are said to be under stress.

Excessively wet years, on the other hand, may cause crop yields to fall. The soil becomes waterlogged and the plant roots rot in the excess water. Intense bursts of rainfall may also damage young plants, both because of the hard impact of the rain drops on the tender plants and because heavy rain can cause soil erosion.

Understanding plants and their environment, and how plants grow and reproduce, is important to effectively manage invasive and native plants. Managers consider this information when developing their IPM plans and selecting control methods.

## PLANTS AND THEIR ENVIRONMENT

The groupings of plants that coexist and interact with one another in an area are called *plant communities*. A plant community is a component of *habitat*, which is a place that provides living organisms with food, water, shelter, and space. Understanding plant communities within the context of their environment is important for managing nonnative plants invading native plant and wildlife habitat.

Although the term plant community generally embraces only plants, plant communities are influenced by the surrounding *ecosystem*. An ecosystem contains all the interactions among living (plants, animals, soil organisms) and nonliving (nutrients, minerals, moisture, disturbances) components. Examples of interactions include nutrient, mineral, and water cycles, and natural disturbances like fire.

The presence of invasive plants, and their management, can affect the interactions of both living and nonliving components of an environment at several levels: plant communities, habitats, and ecosystems. For example, the invasive plant saltcedar (*Tamarix* spp.), which dominates many riparian habitats in the southwestern United States, can greatly affect the water cycle. Dense stands of saltcedar trees can dry up pools, springs, and streams. As saltcedar stands replace native vegetation along waterways, their roots trap sediment, causing the stream channel to narrow. As a result, the flow rate and depth of the water may increase, as well as flooding.

Because of the interconnectedness of the environment, it is important to consider how managing invasive plants will potentially affect other parts of an ecosystem. In some areas of the southwest, saltcedar has replaced native vegetation used for habitat by an endangered bird, the southwestern willow flycatcher (*Empidonax traillii*). Because the bird now uses the invasive trees for habitat, managers are concerned that completely removing saltcedar will further threaten the willow flycatcher. Reestablishing desired plant communities when invasive plants are removed is an important aspect of management.

## PLANT SYSTEMS AND GROWTH FORMS

All plants—native and invasive—share some of the same fundamental characteristics. Understanding how plants function and recognizing their growth forms can help focus control efforts when and where invasive plants are most vulnerable.

### Plant Systems and their Functions

Vascular plants, which include just about all large land and aquatic plants such as trees and shrubs, flowering plants, and ferns, have three systems: vascular system, shoot system, and root system. These systems serve important functions in the processes required for plants to survive and reproduce.

Invasive plant management methods often disrupt processes or inhibit the function of various plant systems to injure or kill the plant. For example, cutting off the shoot system of a plant can prevent it from making food through photosynthesis, or producing flowers and seeds. Biological control agents such as insects or diseases may be released to control invasive plants. They may feed upon, bore through, or infect plant roots, thus reducing the plant's ability to absorb water and nutrients from the soil. Though the mechanisms are often complex, some herbicides work by disrupting vital plant processes such as photosynthesis.

## PLANT GROWTH FORMS

Plant stems, leaves, roots, flowers, cones, and seeds help distinguish one plant from another and may provide information about how a particular plant grows and reproduces. Based on the characteristics of their roots and shoots, plants can be grouped into six general growth forms.

### Plant Reproduction and Dispersal

In general, plants increase their numbers and distribution by reproducing from vegetative structures or by producing seeds or spores. A number of plants are able to reproduce both by seed and by vegetative means. Understanding how plants reproduce and spread aids managers in developing more effective prevention strategies and control methods.

Vegetative reproduction is a process by which new plants are produced from plant parts other than seeds or spores. It is both a natural process in many plant species and one used by horticulturists to start new plants from existing plants. Examples of vegetative reproductive structures are rhizomes (underground stems) and stolons (stems running along the surface of the ground) from which new plants develop.

Plants that spread vegetatively by rhizomes and stolons may produce dense, localized infestations. They may also disperse greater distances if their vegetative structures break off and fall into moving water or are carried to new locations by animals or people. Plants that are able to reproduce vegetatively can present complex management challenges. For example, yellow toadflax (*Linaria vulgaris*) plants produce rhizomes, so pulling toadflax is often ineffective because new plants can sprout from rhizome fragments left in the soil.

### Seeds

Seeds are another way plants reproduce. Seeds are sometimes contained within fruits, which are formed after flowers are pollinated. Plants with showy, fragrant flowers attract insects or birds to serve as pollinators. Plants that do not have flowers (such as pine trees) produce seeds in cones and are pollinated by the wind. Seeds are uniquely designed for various means of dispersal. Some seeds are light with feathery structures that allow them to be carried long distances by the wind. Other seeds are prickly, sticky, or tasty so that they will be transported by attaching to an animal's fur or feet, or passing through the digestive tract. Plants that grow near or in water may have seeds that float along until they land on the bank or shore, or sink to the bottom to germinate.

People also play a role in spreading seeds both accidentally and intentionally. Seed characteristics and dispersal methods present a particular challenge for managing invasive plants like Scotch broom (*Cytisus scoparius*). Scotch broom seed pods typically burst apart into spiral halves, ejecting seeds only a short distance from the parent plant. However, seeds disperse to greater distances with water, soil movement, vehicle tires, human activities, and animals. Even if Scotch broom plants are controlled, seeds in the soil may survive for up to 30 years.

## PLANT LIFE CYCLES AND STRATEGIES

A manager's knowledge of the processes by which a plant completes its life cycle (germinates, grows, reproduces, and dies) can help identify vulnerable life stages and determine the best timing for control. It is also important to know when native or desired plants are vulnerable so that control methods can be used in a way that will not cause injury.

### Plant Life Cycles

Learning to recognize plants when they are in different stages of their life cycle can help with identification for surveying, mapping, or control. For example, if management methods target plants before they flower, it is necessary to be able to identify plants when they are in the vegetative growth stage. Plant life cycles vary, but there are some general phases that most plants—native or nonnative—go through.

## Life History Strategies

The life history of a plant refers to the period in which the life cycle is completed. Some plants grow rapidly, produce seeds, and die in one season, whereas others may live for several years and produce seeds many times. Depending on the amount of time required to complete their life cycle, plants are categorized by three life history strategies: annual (one year), biennial (two years), and perennial (many years).

The types of plant life history strategies are associated with different growth forms. For example, all woody plants (trees, shrubs, and vines) are perennial; grasses are either annual or perennial; and herbaceous forbs can be annual, biennial, and/or perennial.

Understanding different growth forms and associated strategies by which plants complete their life cycle can help managers select and time control methods that target invasive plants when and where they are most vulnerable.

## Annuals

Annuals live for one year and their primary strategy for long-term survival is producing large numbers of seeds or seeds that survive for several years before they germinate.

Unlike deep-rooted perennials, shallow-rooted annual grasses such as Japanese stilt grass (*Microstegium vimineum*) are susceptible to manually pulling as a control method. If these plants have been at the site for years, there are likely to be substantial reserves of seeds in the soil.

## Biennials

Biennials usually take two years to complete their life cycle. During the first growing season, biennials grow leaves, stems, and roots and then enter a period of dormancy. The next spring, biennials grow rapidly, producing flowers, fruits, and seeds before they finally die after the second growth season.

Biennial invasive plants are often targeted for control during their first year of development because they have not yet produced and dispersed their seeds and are more susceptible to some methods, such as herbicides. Houndstongue (*Cynoglossum officinale*), for example, has a stout taproot that is difficult to pull out by hand. Even when the top portion of the plant is removed, biennials like houndstongue can initiate growth from the top of the root, so effective control can require using a method that also damages the root below the soil.

## Perennials

Perennials can live for many years, generating more vegetative growth and creating a stronger plant that can withstand environmental stresses (and control methods). Although perennials may not produce as many seeds as annuals and biennials, they produce seeds year after year. Often their fewer seeds are more likely to germinate and develop into new plants than the seeds of annuals or biennials.

Perennial invasive plants are usually more difficult to control than annuals or biennials because they spread both vegetatively and by dispersing seeds. Stems and roots store food and produce new shoots whenever top growth is killed. Complete control of perennial invasive plants requires killing all plant parts that are capable of producing new shoots or seeds.

# 17

# Flowering Plant

## INTRODUCTION

The flowering plants or angiosperms (Angiospermae or Magnoliophyta) are the most widespread group of land plants. The flowering plants and the gymnosperms are the only extant groups of seed plants. The flowering plants are distinguished from other seed plants by a series of apomorphies, or derived characteristics.

The flowers, which are the reproductive organs of flowering plants, are the most remarkable feature distinguishing them from other seed plants. Flowers aided angiosperms by enabling a wider range of evolutionary relationships and broadening the ecological niches open to them, allowing flowering plants to eventually dominate terrestrial ecosystems.

### Stamens with Two Pairs of Pollen Sacs

Stamens are much lighter than the corresponding organs of gymnosperms and have contributed to the diversification of angiosperms through time with adaptations to specialized pollination syndromes, such as particular pollinators. Stamens have also become modified through time to prevent self-fertilization, which has permitted further diversification, allowing angiosperms eventually to fill more niches.

### Reduced Male Parts, Three Cells

The male gametophyte in angiosperms is significantly reduced in size compared to those of gymnosperm seed plants. The smaller pollen decreases the time from pollination – the pollen grain reaching the female plant – to fertilization of the ovary; in gymnosperms fertilization can occur up to a year after pollination, while in angiosperms the fertilization begins very soon after pollination. The shorter time leads to angiosperm plants setting seeds sooner and faster than gymnosperms, which is a distinct evolutionary advantage.

The closed carpel of angiosperms also allows adaptations to specialized pollination syndromes and controls to prevent self-fertilization, thereby maintaining increased diversity. Once the ovary is fertilized the carpel and some surrounding tissues develop into a fruit, another opportunity for angiosperms to increase their domination of the terrestrial ecosystem with evolutionary adaptations to dispersal mechanisms.

### Reduced Female Gametophyte, Seven Cells with Eight Nuclei

The reduced female gametophyte, like the reduced male gametophyte may be adaptations allowing for more rapid seed set, eventually leading to such flowering plant adaptations as annual herbaceous life cycles, allowing the flowering plants to fill even more niches.

### Endosperm

Endosperm formation generally begins after fertilization and before the first division of the zygote. Endosperm is a highly nutritive tissue that can provide food for the developing embryo, the cotyledons, and sometimes for the seedling when it first appears.

These distinguishing characteristics taken together have made the angiosperms the most diverse and numerous land plants and the most commercially important group to humans. The major exception to the dominance of terrestrial ecosystems by flowering plants is the coniferous forest.

## EVOLUTION

Land plants have existed for about 425 million years. Early land plants reproduced by spores like their aquatic counterparts. Marine organisms can easily scatter copies of themselves to float away and grow elsewhere. Land plants soon found it advantageous to protect their copies from drying out and other hazards by enclosing them in a case, the seed. Early seed bearing plants, like the ginkgo, and conifers (such as pines and firs), did not produce flowers.

While there is only hard evidence of such flowers existing about 130 million years ago, there is some circumstantial evidence that they may have existed 250 million years ago. A chemical used by plants to defend their flowers, oleanane, has been detected in fossil plants that old, including gigantopterids which evolved at that time and bear many of the traits of modern, flowering plants, though they are not known to be flowering plants themselves, because only their stems and prickles have been found preserved in detail, one of the earliest examples of petrification.

The apparently sudden appearance of relatively modern flowers in the fossil record posed such a problem for the theory of evolution that it was called an "abominable mystery" by Charles Darwin However, the fossil record has grown since the time of Darwin, and recently discovered angiosperm fossils such as *Archaefructus*, along with further discoveries of fossil gymnosperms, suggest how angiosperm characteristics may have been acquired in a series of steps. Several groups of extinct gymnosperms, particularly seed ferns, have been proposed as the ancestors of flowering plants but there is no continuous fossil evidence showing exactly how flowers evolved. Some older fossils, such as the upper Triassic *Sanmiguelia*, have been suggested. Based on current evidence, some propose that the ancestors of the angiosperms diverged from an unknown group of gymnosperms during the late Triassic (245-202 million years ago). The relationship of the earlier gigantopterids to flowering plants is still enigmatic.

A close relationship between Angiosperms and Gnetophytes, suggested on the basis of morphological evidence, has been disputed on the basis of molecular evidence that suggest Gnetophytes are more closely related to other gymnosperms.

Recent DNA analysis (molecular systematics) show that *Amborella trichopoda,* found on the Pacific island of New Caledonia, belongs to a sister group of the other flowering plants, and morphological studies suggest that it has features which may have been characteristic of the earliest flowering plants.

The great angiosperm radiation, when a great diversity of angiosperms appears in the fossil record, occurred in the mid-Cretaceous (approximately 100 million years ago). However, a study in 2007 estimated that the division of the five most recent (the genus *Ceratophyllum*, the family Chloranthaceae, the eudicots, the magnoliids, and the monocots) of the eight main groups occurred around 140 million years ago. By the late Cretaceous, angiosperms appear to have become the predominant group of land plants, and many fossil plants recognizable as belonging to modern families (including beech, oak, maple, and magnolia) appeared. However, some authors have proposed an earlier origin for angiosperms, sometime in the Paleozoic (251 million years ago or more).

It is generally assumed that the function of flowers, from the start, was to involve mobile animals in their reproduction processes. Pollen can be scattered without bright colors and obvious shapes. Expending energy on these structures would appear to be a liability, unless they provide significant benefit.

Island genetics provides one proposed explanation for the sudden, fully developed appearance of flowering plants. Island genetics is believed to be a common source of speciation in general, especially when it comes to radical adaptations which seem to have required inferior transitional forms. Flowering plants may have evolved in an isolated setting like an island or island chain, where the plants bearing them were able to develop a highly specialized relationship with some specific animal (a wasp, for example). Such a relationship, with a hypothetical wasp carrying pollen from one plant to another much the way fig wasps do today, could result in both the plant(s) and their partners developing a high degree of specialization. Note that the wasp example is not incidental; bees, which apparently evolved specifically due to mutualistic plant relationships, are descended from wasps.

Animals are also involved in the distribution of seeds. Fruit, which is formed by the enlargement flower parts, is frequently a seed disbursal tool which depends upon animals, which eat or otherwise disturb it, incidentally scattering the seeds it contains (see frugivory). While many such mutualistic relationships remain too fragile to survive competition with mainland animals and spread, flowers proved to be an unusually effective means of production, spreading (whatever their actual origin) to become the dominant form of land plant life.

Flowers are derived from leaf and stem components, arising from a combination of genes normally responsible for forming new shoots. The most primitive flowers are thought to have had a variable number of flower parts, often separate from (but in contact with) each other. The flowers would have tended to grow in a spiral pattern, to be bisexual (in plants, this means both male and female parts on the same flower), and to be dominated by the ovary (female part). As flowers grew more advanced, some variations developed parts fused together, with a much more specific number and design, and with either specific sexes per flower or plant, or at least "ovary inferior".

Flower evolution continues to the present day; modern flowers have been so profoundly influenced by humans that some of them cannot be pollinated in nature. Many modern, domesticated flowers used to be simple weeds, which only sprouted when the ground was disturbed. Some of them tended to grow with human crops, perhaps already having symbiotic companion plant relationships with them, and the prettiest did not get plucked because of their beauty, developing a dependence upon and special adaptation to human affection

## CLASSIFICATION

There are eight groups of living angiosperms:

- *Amborella*- a single species of shrub from New Caledonia
- *Nymphaeales*- about 80 species-water lilies and Hydatellaceae

- *Austrobaileyales*- about 100 species of woody plants from various parts of the world.
- *Chloranthales*- several dozen species of aromatic plants with toothed leaves.
- *Ceratophyllum*- about 6 species of aquatic plants, perhaps most familiar as aquarium plants.
- *magnoliids*- about 9,000 species, characterized by trimerous flowers, pollen with one pore, and usually branching-veined leaves - for example magnolias, bay laurel, and black pepper.
- *eudicots*- about 175,000 species, characterized by 4- or 5-merous flowers, pollen with three pores, and usually branching-veined leaves - for example sunflowers, petunia, buttercup, apples and oaks.
- *monocots*- about 70,000 species, characterized by trimerous flowers, a single cotyledon, pollen with one pore, and usually parallel-veined leaves - for example grasses, orchids, and palms.

The exact relationship between these eight groups is not yet clear, although it has been determined that the first three groups to diverge from the ancestral angiosperm were Amborellales, Nymphaeales, and Austrobaileyales, in that order.

## History of Classification

The botanical term "Angiosperm", from the Ancient Greek , *angeíon* (receptacle, vessel) and *spérma* (seed), was coined in the form Angiospermae by Paul Hermann in 1690, as the name of that one of his primary divisions of the plant kingdom. This included flowering plants possessing seeds enclosed in capsules, distinguished from his Gymnospermae, or flowering plants with achenial or schizo-carpic fruits, the whole fruit or each of its pieces being here regarded as a seed and naked. The term and its antonym were maintained by Carolus Linnaeus with the same sense, but with restricted application, in the names of the orders of his class Didynamia. Its use with any approach to its modern scope only became possible after 1827, when Robert Brown established the existence of truly naked ovules in the Cycadeae and Coniferae, and applied to them the name Gymnosperms. From that time onwards, so long as these Gymnosperms were, as was usual, reckoned as dicotyledonous flowering plants, the term Angiosperm was used antithetically by botanical writers, with varying scope, as a group-name for other dicotyledonous plants.

In 1851, Hofmeister discovered the changes occurring in the embryo-sac of flowering plants, and determined the correct relationships of these to the Cryptogamia. This fixed the position of Gymnosperms as a class distinct from Dicotyledons, and the term Angiosperm then gradually came to be accepted as the suitable designation for the whole of the flowering plants other than Gymnosperms, including the classes of Dicotyledons and Monocotyledons. This is the sense in which the term is used today.

In most taxonomies, the flowering plants are treated as a coherent group. The most popular descriptive name has been Angiospermae (Angiosperms), with Anthophyta ("flowering plants") a second choice. These names are not linked to any rank. The Wettstein system and the Engler system use the name Angiospermae, at the assigned rank of subdivision. The Reveal system treated flowering plants as subdivision Magnoliophytina but later split it to Magnoliopsida, Liliopsida and Rosopsida. The Takhtajan system and Cronquist system treat this group at the rank of division, leading to the name Magnoliophyta (from the family name Magnoliaceae). The Dahlgren system and Thorne system (1992) treat this group at the rank of class, leading to the name Magnoliopsida. However, the APG system, of 1998, and the APG II system, of 2003 do not treat it as a formal taxon but rather treat it as a clade without a formal botanical name and use the name angiosperms for this clade.

The internal classification of this group has undergone considerable revision. The Cronquist system, proposed by Arthur Cronquist in 1968 and published in its full form in 1981, is still widely used, but is no longer believed to accurately reflect phylogeny. A general consensus about how the flowering plants should be arranged has recently begun to emerge, through the work of the Angiosperm Phylogeny Group, who published an influential reclassification of the angiosperms in 1998. An update incorporating more recent research was published as APG II in 2003.

Traditionally, the flowering plants are divided into two groups, which in the Cronquist system are called *Magnoliopsida* (at the rank of class, formed from the family name *Magnoliacae*) and *Liliopsida* (at the rank of class, formed from the family name *Liliaceae*). Other descriptive names allowed by Article 16 of the ICBN include *Dicotyledones* or *Dicotyledoneae*, and *Monocotyledones* or *Monocotyledoneae*, which have a long history of use. In English a member of either group may be called a *dicotyledon* (plural *dicotyledons*) and *monocotyledon* (plural *monocotyledons*), or abbreviated, as *dicot* (plural *dicots*) and *monocot* (plural *monocots*). These names derive from the observation that the dicots most often have two *cotyledons*, or embryonic leaves, within each seed. The monocots usually have only one, but the rule is not absolute either way. From a diagnostic point of view the number of cotyledons is neither a particularly handy nor reliable character.

Recent studies, as by the APG, show that the monocots form holophyletic or monophyletic group; this clade is given the name *monocots*. However, the dicots are not (they are a paraphyletic group). Nevertheless, within the dicots a monophyletic group does exist, called the *eudicots* or *tricolpates*, and including most of the dicots. The name *tricolpates* derives from a type of pollen found widely within this group. The name *eudicots* is formed combining *dicot* with the prefix *eu-* (from Greek, for "well," or "good," botanically indicating "true"), as the eudicots share the characters traditionally attributed to the dicots, such as flowers with four or five parts (four or five petals, four or five sepals). Separating this group of eudicots from the rest of the (former) dicots leaves a remainder, which sometimes are called informally *palaeodicots* (Greek prefix "*palaeo-*" means "old"). As this remnant group is not monophyletic this is a term of convenience only.

## FLOWERING PLANT DIVERSITY

The number of species of flowering plants is estimated to be in the range of 250,000 to 400,000. The number of families in APG (1998) was 462. In APG II (2003) it is not settled; at maximum it is 457, but within this number there are 55 optional segregates, so that the minimum number of families in this system is 402.

The diversity of flowering plants is not evenly distributed. Nearly all species belong to the eudicot (75%), monocot (23%) and magnoliid (2%) clades. The remaining 5 clades contain a little over 250 species in total, i.e., less than 0.1% of flowering plant diversity, divided among 9 families.

The most diverse families of flowering plants, in their APG circumscriptions, in order of number of species, are:

| | |
|---|---|
| Asteraceae or Compositae (daisy family): | 23,600 species |
| Orchidaceae (orchid family): | 21,950 species |
| Fabaceae or Leguminosae (pea family): | 19,400 |
| Rubiaceae (madder family): | 13,183 |
| Poaceae or Gramineae (grass family): | 10,035 |
| Lamiaceae or Labiatae (mint family): | 7,173 |
| Euphorbiaceae (spurge family): | 5,735 |
| Cyperaceae (sedge family): | 4,350 |
| Malvaceae (mallow family): | 4,225 |
| Araceae (aroid family): | 4,025 |

In the list above (showing only the 10 largest families), the Orchidaceae, Poaceae, Cyperaceae and Araceae are monocot families; the others are dicot families.

## VASCULAR ANATOMY

The amount and complexity of tissue-formation in flowering plants exceeds that of Gymnosperms. The vascular bundles of the stem are arranged such that the xylem and phloem form concentric rings.

In the Dicotyledons, the bundles in the very young stem are arranged in an open ring, separating a central pith from an outer cortex. In each bundle, separating the xylem and phloem, is a layer of meristem or active formative tissue known as cambium; by the formation of a layer of cambium between the bundles (interfascicular cambium) a complete ring is formed, and a regular periodical increase in thickness results from the development of xylem on the inside and phloem on the outside. The soft phloem becomes crushed, but the hard wood persists and forms the bulk of the stem and branches of the woody perennial. Owing to differences in the character of the elements produced at the beginning and end of the season, the wood is marked out in transverse section into concentric rings, one for each season of growth, called annual rings.

Among the Monocotyledons, the bundles are more numerous in the young stem and are scattered through the ground tissue. They contain no cambium and once formed the stem increases in diameter only in exceptional cases.

The characteristic feature of angiosperms is the flower. Flowers show remarkable variation in form and elaboration, and provide the most trustworthy external characteristics for establishing relationships among angiosperm species. The function of the flower is to ensure fertilization of the ovule and development of fruit containing seeds. The floral apparatus may arise terminally on a shoot or from the axil of a leaf. Occasionally, as in violets, a flower arises singly in the axil of an ordinary foliage-leaf. More typically, the flower-bearing portion of the plant is sharply distinguished from the foliage-bearing or vegetative portion, and forms a more or less elaborate branch-system called an inflorescence.

The reproductive cells produced by flowers are of two kinds. Microspores which will divide to become pollen grains, are the "male" cells and are borne in the stamens (or microsporophylls). The "female" cells called megaspores, which will divide to become the egg-cell (megagametogenesis), are contained in the ovule and enclosed in the carpel (or megasporophyll).

The flower may consist only of these parts, as in willow, where each flower comprises only a few stamens or two carpels. Usually other structures are present and serve to protect the sporophylls and to form an envelope attractive to pollinators. The individual members of these surrounding structures are known as sepals and petals (or tepals in flowers such as *Magnolia* where sepals and petals are not distinguishable from each other). The outer series (calyx of sepals) is usually green and leaf-like, and functions to protect the rest of the flower, especially the bud. The inner series (corolla of petals) is generally white or brightly colored, and is more delicate in structure. It functions to attract insect or bird pollinators. Attraction is effected by color, scent, and nectar, which may be secreted in some part of the flower. The characteristics that attract pollinators account for the popularity of flowers and flowering plants among humans.

While the majority of flowers are perfect or hermaphrodite (having both male and female parts in the same flower structure), flowering plants have developed numerous morphological and physiological mechanisms to reduce or prevent self-fertilization. Heteromorphic flowers have short carpels and long stamens, or vice versa, so animal pollinators cannot easily transfer pollen to the pistil (receptive part of the carpel). Homomorphic flowers may employ a biochemical (physiological) mechanism called self-incompatibility to discriminate between self- and non-self pollen grains. In other species, the male and female parts are morphologically separated, developing on different flowers.

Double fertilization refers to a process in which two sperm cells fertilize two cells in the ovary. The pollen grain adheres to the stigma of the carpel (female reproductive structure) and grows a pollen tube that penetrates the ovum through a tiny pore called a micropyle. Two sperm cells are released into the ovary through this tube. One of the two sperm cells fertilizes the egg cell, forming a diploid zygote or embryo, also called the ovule. The other sperm cell fuses with two haploid polar nuclei in the center of the embryo sac. The resulting cell is triploid (3n). This triploid cell divides through mitosis and forms the endosperm, a nutrient-rich tissue inside the fruit. When seed develops without fertilization, the process is known as apomixis.

As the development of embryo and endosperm proceeds within the embryo-sac, the sac wall enlarges and combines with the nucellus (which is likewise enlarging) and the integument to form the *seed-coat*. The ovary wall develops to form the fruit or pericarp, whose form is closely associated with the manner of distribution of the seed. Frequently the influence of fertilization is felt beyond the ovary, and other parts of the flower take part in the formation of the fruit, e.g. the floral receptacle in the apple, strawberry and others.

The character of the seed-coat bears a definite relation to that of the fruit. They protect the embryo and aid in dissemination; they may also directly promote germination. Among plants with indehiscent fruits, the fruit generally provides protection for of the embryo and secures dissemination. In this case, the seed-coat is only slightly developed. If the fruit is dehiscent and the seed is exposed, the seed-coat is generally well developed, and must discharge the functions otherwise executed by the fruit.

## ECONOMIC IMPORTANCE

Agriculture is almost entirely dependent on angiosperms, either directly or indirectly through livestock feed. Of all the families plants, the Poaceae, or grass family, is by far the most important, providing the bulk of all feedstocks (rice, corn (maize), wheat, barley, rye, oats, pearl millet, sugar cane, sorghum). The Fabaceae, or legume family, comes in second place. Also of high importance are the Solanaceae, or nightshade family (potatoes, tomatoes, and peppers, among others), the Cucurbitaceae, or gourd family (also including pumpkins and melons), the Brassicaceae, or mustard plant family (including rapeseed and cabbage), and the Apiaceae, or parsley family. Many of our fruits come from the Rutaceae, or rue family, and the Rosaceae, or rose family (including apples, pears, cherries, apricots, plums, etc).

In some parts of the world, certain single species assume paramount importance because of their variety of uses, for example the coconut (*Cocos nucifera*) on Pacific atolls, and the olive (*Olea europaea*) in the Mediterranean region.

Flowering plants also provide economic resources in the form of wood, paper, fiber (cotton, flax, and hemp, among others), medicines (digitalis, camphor), decorative and landscaping plants, and many other uses. The main area in which they are surpassed by other plants is timber production.

# Index

D



E

## J



## K



## L



## M



## N



## O



## P



## Q

## R



## S



## T

❑❑❑